SPSS® 12.0 Brief Guide

For more information about SPSS® software products, please visit our Web site at *http://www.spss.com* or contact

SPSS Inc.
233 South Wacker Drive, 11th Floor
Chicago, IL 60606-6412
Tel: (312) 651-3000
Fax: (312) 651-3668

SPSS is a registered trademark and the other product names are the trademarks of SPSS Inc. for its proprietary computer software. No material describing such software may be produced or distributed without the written permission of the owners of the trademark and license rights in the software and the copyrights in the published materials.

The SOFTWARE and documentation are provided with RESTRICTED RIGHTS. Use, duplication, or disclosure by the Government is subject to restrictions as set forth in subdivision (c) (1) (ii) of The Rights in Technical Data and Computer Software clause at 52.227-7013. Contractor/manufacturer is SPSS Inc., 233 South Wacker Drive, 11th Floor, Chicago, IL 60606-6412.

General notice: Other product names mentioned herein are used for identification purposes only and may be trademarks of their respective companies.

TableLook is a trademark of SPSS Inc.
Windows is a registered trademark of Microsoft Corporation.
DataDirect, DataDirect Connect, INTERSOLV, and SequeLink are registered trademarks of DataDirect Technologies.
Portions of this product were created using LEADTOOLS © 1991-2000, LEAD Technologies, Inc. ALL RIGHTS RESERVED.
LEAD, LEADTOOLS, and LEADVIEW are registered trademarks of LEAD Technologies, Inc.
Portions of this product were based on the work of the FreeType Team (*http://www.freetype.org*).

SPSS® 12.0 Brief Guide
Copyright © 2003 by SPSS Inc.
All rights reserved.
Printed in the United States of America.

No part of this publication may be reproduced, stored in a retrieval system, or transmitted, in any form or by any means, electronic, mechanical, photocopying, recording, or otherwise, without the prior written permission of the publisher.

3 4 5 6 7 8 9 0   06 05 04 03

ISBN 0-13-109675-3

Preface

The *SPSS® 12.0 Brief Guide* provides a set of tutorials designed to acquaint you with the various components of the SPSS system. You can work through the tutorials in sequence or turn to the topics for which you need additional information. You can use this book as a supplement to the online tutorial that is included with the SPSS Base 12.0 system or ignore the online tutorial and start with the tutorials found here.

SPSS 12.0

SPSS 12.0 is a comprehensive system for analyzing data. SPSS can take data from almost any type of file and use them to generate tabulated reports, charts, and plots of distributions and trends, descriptive statistics, and complex statistical analyses.

SPSS makes statistical analysis more accessible for the beginner and more convenient for the experienced user. Simple menus and dialog box selections make it possible to perform complex analyses without typing a single line of command syntax. The Data Editor offers a simple and efficient spreadsheet-like facility for entering data and browsing the working data file.

Internet Resources

The SPSS Web site (*http://www.spss.com*) offers answers to frequently asked questions about installing and running SPSS software and provides access to data files and other useful information.

In addition, the SPSS USENET discussion group (not sponsored by SPSS) is open to anyone interested in SPSS products. The USENET address is *comp.soft-sys.stat.spss*. It deals with computer, statistical, and other operational issues related to SPSS software.

You can also subscribe to an e-mail message list that is gatewayed to the USENET group. To subscribe, send an e-mail message to *listserv@listserv.uga.edu*. The text of the e-mail message should be: subscribe SPSSX-L firstname lastname. You can then post messages to the list by sending an e-mail message to *listserv@listserv.uga.edu*.

Additional Publications

For additional information about the features and operations of SPSS Base 12.0, you can consult the *SPSS Base 12.0 User's Guide*, which includes information on standard graphics. Complete information about using interactive graphics can be found in *SPSS Interactive Graphics 10.0*, which is compatible with release 12.0 of SPSS. Examples using the statistical procedures found in SPSS Base 12.0 are provided in the Help system, installed with the software. Algorithms used in the statistical procedures are available on the product CD-ROM.

In addition, beneath the menus and dialog boxes, SPSS uses a command language. Some extended features of the system can be accessed only via command syntax. (Those features are not available in the Student Version.) Complete command syntax is documented in the *SPSS 12.0 Command Syntax Reference*, provided on the product CD-ROM.

Individuals worldwide can order additional product manuals directly from SPSS Inc. For telephone orders in the United States and Canada, call SPSS Inc. at 800-543-2185. For telephone orders outside of North America, contact your local office, listed on the SPSS Web site at *http://www.spss.com/worldwide*.

The *SPSS Statistical Procedures Companion*, by Marija Norušis, is being prepared for publication by Prentice Hall. It contains overviews of the procedures in the SPSS Base, plus Logistic Regression, General Linear Models, and Linear Mixed Models. Further information will be available on the SPSS Web site at *http://www.spss.com* (click Store, select your country, and click Books).

SPSS Options

The following options are available as add-on enhancements to the full (not Student Version) SPSS Base system:

SPSS Regression Models™ provides techniques for analyzing data that do not fit traditional linear statistical models. It includes procedures for probit analysis, logistic regression, weight estimation, two-stage least-squares regression, and general nonlinear regression.

SPSS Advanced Models™ focuses on techniques often used in sophisticated experimental and biomedical research. It includes procedures for general linear models (GLM), linear mixed models, variance components analysis, loglinear analysis, ordinal regression, actuarial life tables, Kaplan-Meier survival analysis, and basic and extended Cox regression.

SPSS Tables™ creates a variety of presentation-quality tabular reports, including complex stub-and-banner tables and displays of multiple response data.

SPSS Trends™ performs comprehensive forecasting and time series analyses with multiple curve-fitting models, smoothing models, and methods for estimating autoregressive functions.

SPSS Categories® performs optimal scaling procedures, including correspondence analysis.

SPSS Conjoint™ performs conjoint analysis.

SPSS CHAID™ simplifies tabular analysis of categorical data, develops predictive models, screens out extraneous predictor variables, and produces easy-to-read tree diagrams that segment a population into subgroups that share similar characteristics.

SPSS Exact Tests™ calculates exact p values for statistical tests when small or very unevenly distributed samples could make the usual tests inaccurate.

SPSS Missing Value Analysis™ describes patterns of missing data, estimates means and other statistics, and imputes values for missing observations.

SPSS Maps™ turns your geographically distributed data into high-quality maps with symbols, colors, bar charts, pie charts, and combinations of themes to present not only what is happening but where it is happening.

SPSS Complex Samples™ allows survey, market, health, and public opinion researchers, as well as social scientists who use sample survey methodology, to incorporate their complex sample designs into data analysis.

The SPSS family of products also includes applications for data entry, text analysis, classification, neural networks, and flowcharting.

Training Seminars

SPSS Inc. provides both public and onsite training seminars for SPSS. All seminars feature hands-on workshops. SPSS seminars will be offered in major U.S. and European cities on a regular basis. For more information on these seminars, contact your local office, listed on the SPSS Web site at *http://www.spss.com/worldwide*.

Technical Support

The services of SPSS Technical Support are available to registered customers of SPSS. (Student Version customers should read the special section on technical support for the Student Version. For more information, see "Technical Support for Students" on page vii.) Customers may contact Technical Support for assistance in using SPSS products or for installation help for one of the supported hardware environments. To reach Technical Support, see the SPSS Web site at *http://www.spss.com*, or contact your local office, listed on the SPSS Web site at *http://www.spss.com/worldwide*. Be prepared to identify yourself, your organization, and the serial number of your system.

Tell Us Your Thoughts

Your comments are important. Please let us know about your experiences with SPSS products. We especially like to hear about new and interesting applications using the SPSS system. Please send e-mail to *suggest@spss.com*, or write to SPSS Inc., Attn: Director of Product Planning, 233 South Wacker Drive, 11th Floor, Chicago IL 60606-6412.

SPSS 12.0 for Windows Student Version

The SPSS 12.0 for Windows Student Version is a limited but still powerful version of the SPSS Base 12.0 system.

Capability

The Student Version contains all of the important data analysis tools contained in the full SPSS Base system, including:

- Spreadsheet-like Data Editor for entering, modifying, and viewing data files.
- Statistical procedures, including t tests, analysis of variance, crosstabulations, and multidimensional scaling.
- Interactive graphics that allow you to change or add chart elements and variables dynamically; the changes appear as soon as they are specified.
- Standard high-resolution graphics for an extensive array of analytical and presentation charts and tables.

Limitations

Created for classroom instruction, the Student Version is limited to use by students and instructors for educational purposes only. The Student Version does not contain all of the functions of the SPSS Base 12.0 system. The following limitations apply to the SPSS 12.0 for Windows Student Version:

- Data files cannot contain more than 50 variables.
- Data files cannot contain more than 1500 cases. SPSS add-on modules (such as Regression Models or Advanced Models) cannot be used with the Student Version.
- SPSS command syntax is not available to the user. This means that it is not possible to repeat an analysis by saving a series of commands in a syntax or "job" file, as can be done in the full version of SPSS.
- Scripting and automation are not available to the user. This means that you cannot create scripts that automate tasks that you repeat often, as can be done in the full version of SPSS.

Technical Support for Students

Students should obtain technical support from their instructors or from local support staff identified by their instructors. Technical support from SPSS for the SPSS 12.0 Student Version is provided *only to instructors using the system for classroom instruction*.

Before seeking assistance from your instructor, please write down the information described below. Without this information, your instructor may be unable to assist you:

- The type of PC you are using, as well as the amount of RAM and free disk space you have.
- The operating system of your PC.
- A clear description of what happened and what you were doing when the problem occurred. If possible, please try to reproduce the problem with one of the sample data files provided with the program.
- The exact wording of any error or warning messages that appeared on your screen.
- How you tried to solve the problem on your own.

Technical Support for Instructors

Instructors using the Student Version for classroom instruction may contact SPSS Technical Support for assistance. In the United States and Canada, call SPSS Technical Support at 312-651-3410, or send an e-mail to *support@spss.com*. Please include your name, title, and academic institution.

Instructors outside of the United States and Canada should contact your local SPSS office, listed on the SPSS Web site at *http://www.spss.com/worldwide*.

Contents

1 Introduction   1

Sample Files 1
Starting SPSS 2
      Variable Display in Dialog Boxes 3
Opening a Data File. 3
Running an Analysis 7
Viewing Results 11
Creating Charts. 12
Exiting SPSS. 14

2 Using the Help System   15

Help Contents Tab. 16
Help Index Tab 17
Dialog Box Help 18
Statistics Coach 19
Case Studies 27

3 Reading Data   29

Basic Structure of an SPSS Data File 29
Reading an SPSS Data File 30
Reading Data from Spreadsheets 31
Reading Data from a Database 34
Reading Data from a Text File 40
Saving Data 48

4 Using the Data Editor   51

Entering Numeric Data 51
Entering String Data 54
Defining Data 56
    Adding a Variable Label 56
    Changing Variable Type and Format 57
    Adding Value Labels for Numeric Variables 58
    Adding Value Labels for String Variables 60
    Using Value Labels for Data Entry 61
    Handling Missing Data 62
    Copying and Pasting Value Attributes 66
    Defining Variable Properties for Categorical Variables 72

5 Examining Summary Statistics for Individual Variables   79

Level of Measurement 79
Summary Measures for Categorical Data 80
    Charts for Categorical Data 81
Summary Measures for Scale Variables 83
    Histograms for Scale Variables 86

6 Working with Output   89

Using the Viewer 89
Using the Pivot Table Editor 92
    Accessing Output Definitions 92
    Pivoting Tables 93

    Creating and Displaying Layers 96
    Editing Tables 98
    Hiding Rows and Columns 100
    Changing Data Display Formats 100
TableLooks 102
    Using Predefined Formats 102
    Customizing TableLook Styles 103
    Changing the Default Table Formats. 105
    Customizing the Initial Display Settings 107
    Displaying Variable and Value Labels. 109
Using Results in Other Applications 111
    Pasting Results as Rich Text 112
    Pasting Results as Metafiles 113
    Pasting Results as Text 115
    Exporting Results to Microsoft Word and Excel Files 117
    Exporting Results to HTML and Text Formats 123

7 Creating and Editing Charts   125

Creating Standard Charts 125
Editing Standard Charts 128
Creating Interactive Charts. 135
Editing Interactive Charts 139
Creating an Interactive Chart from a Pivot Table 143

8 Working with Syntax   147

Pasting Syntax 147
Editing Syntax. 150
Typing Syntax 151

Saving Syntax. 152
Opening and Running a Syntax File 152

9 Modifying Data Values   153

Creating a Categorical Variable from a Scale Variable. 153
Computing New Variables. 159
    Using Functions in Expressions 161
    Using Conditional Expressions 162
Computing Variables with Missing Values. 164

10 Sorting and Selecting Data   169

Sorting Data 169
Split-File Processing. 170
    Sorting Cases for Split-File Processing 173
    Turning Split-File Processing On and Off 173
Selecting Subsets of Cases. 174
    Selecting Cases Based on Conditional Expressions 175
    Selecting a Random Sample 175
    Selecting a Time Range or Case Range 176
    Unselected Cases 178
Case Selection Status. 178

11 Additional Statistical Procedures   181

Summarizing Data. 181
    Frequencies. 182

Explore	184
More about Summarizing Data	186
Comparing Means	**186**
Means	187
Paired-Samples T Test	188
More about Comparing Means	189
ANOVA Models	**190**
Univariate Analysis of Variance	190
Correlating Variables	**192**
Bivariate Correlations	192
Partial Correlations	192
Regression Analysis	**193**
Linear Regression	194
Nonparametric Tests	**195**
Chi-Square	195
Time Series Analysis	**197**
Exponential Smoothing	198

Index *203*

Chapter 1

Introduction

This guide provides a set of tutorials designed to acquaint you with the various components of the SPSS system. You can work through the tutorials in sequence or turn to the topics for which you need additional information. The goal is to enable you to perform useful analyses on your data with SPSS.

This chapter will introduce you to the basic environment of SPSS and demonstrate a typical session. We will run SPSS, retrieve a previously defined SPSS data file, and then produce a simple statistical summary and a chart. In the process, you will learn the roles of the primary windows within SPSS and see a few features that smooth the way when running analyses.

More detailed instruction about many of the topics touched upon in this chapter will follow in later chapters. Here, we hope to give you a basic framework for understanding and using SPSS.

Sample Files

Most of the examples presented here use the data file *demo.sav*. This data file is a fictitious survey of several thousand people, containing basic demographic and consumer information.

All sample files used in these examples are located in the folder in which SPSS is installed or in the *tutorial\sample_files* folder within the SPSS installation folder.

Chapter 1

Starting SPSS

To start SPSS:

▶ From the Windows Start menu choose:
Programs
  SPSS for Windows
    SPSS for Windows

To start SPSS for Windows Student Version:

▶ From the Windows Start menu choose:
Programs
  SPSS for Windows Student Version

When you start a session, you see the Data Editor window.

Figure 1-1
Data Editor window (Data View)

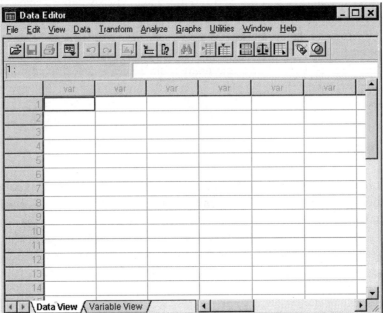

Introduction

Variable Display in Dialog Boxes

Either variable names or longer variable labels will appear in list boxes in dialog boxes. Additionally, variables in list boxes can be ordered alphabetically or by their position in the file.

In this guide, we will display variable labels in alphabetical order within list boxes. For a new user of SPSS, this provides a more complete description of variables in an easy-to-follow order.

Since the default setting within SPSS is to display variable labels in file order, we will change this before accessing data.

▶ From the menus choose:
Edit
  Options...

▶ Select Display labels in the Variable Lists group on the General tab.

▶ Also select Alphabetical.

▶ Click OK, and then click OK to confirm the change.

Opening a Data File

Before you can analyze data, you need some data to analyze.

To open a data file:

▶ From the menus choose:
File
  Open
    Data...

Alternatively, you can use the Open File button on the toolbar.

Figure 1-2
Open File toolbar button

This opens the Open File dialog box.

Figure 1-3
Open File dialog box

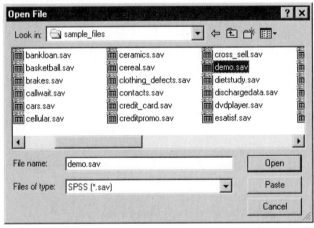

By default, SPSS-format data files (*.sav* extension) are displayed. You can display other file formats using the Files of Type drop-down list.

By default, data files in the folder (directory) in which SPSS is installed are displayed. The files for this guide are located in the folder in which SPSS is installed or in the *tutorial\sample_files* folder within the SPSS installation folder.

► Double-click the *tutorial* folder.

► Double-click the *sample_files* folder.

Introduction

▶ Click the file *demo.sav* in the files list box (or just *demo* if file extensions are not displayed).

Figure 1-4
Sample_files folder displayed in Open File dialog box

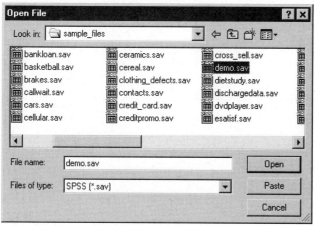

▶ Click Open to open the SPSS data file.

Figure 1-5
Variable labels

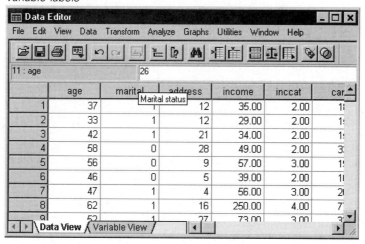

The data file is displayed in the Data Editor. If you put the mouse cursor on a variable name (the column headings), a more descriptive variable label is displayed if one has been defined for that variable.

By default, the actual data values are displayed. To display labels:

▶ From the menus choose:
View
  Value Labels

Alternatively, you can use the Label tool on the toolbar.

Figure 1-6
Value Labels tool

Descriptive value labels are now displayed. This makes it easier to interpret the responses.

Figure 1-7
Value labels displayed in the Data Editor

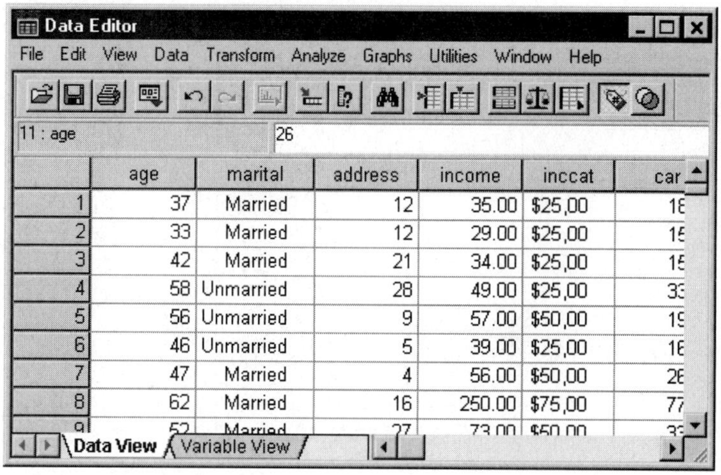

Running an Analysis

The Analyze menu contains a list of general reporting and statistical analysis categories. Most of the categories are followed by an arrow, which indicates that there are several analysis procedures available within the category; they will appear on a submenu when the category is selected.

We'll start with a simple frequency table (table of counts).

▶ From the menus choose:

Analyze
  Descriptive Statistics
    Frequencies...

This opens the Frequencies dialog box.

Figure 1-8
Frequencies dialog box

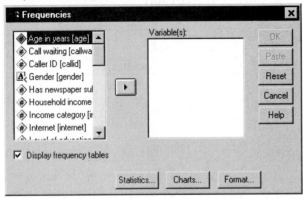

▶ Select (click) the variable *Income category*.

Figure 1-9
Variable labels and names in the Frequencies dialog box

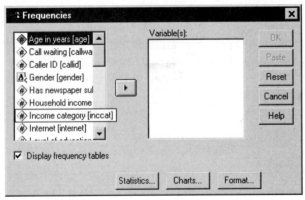

A more complete description of each variable pops up when the cursor is over it. The variable name (in square brackets) is *inccat*, and it has the variable label *Income category*. If there were no variable label, only the variable name would appear in the list box.

Introduction

In the dialog box, you choose the variables you want to analyze from the source list on the left and move them into the Variable(s) list on the right. The OK button, which runs the analysis, is disabled until at least one variable is placed in the Variable(s) list.

Additional labeling information can be easily obtained for any variable on the list by clicking on the variable name with the right mouse button.

▶ Click the right mouse button on *Income Category [inccat]*, and then click (left mouse button) Variable Information.

▶ Click the down arrow on the Value Labels drop-down list.

Figure 1-10
Defined labels for income variable

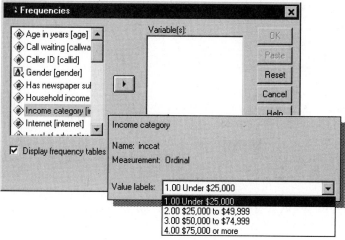

All of the defined value labels for the variable are displayed.

▶ Click *Gender [gender]* in the source variable list, and then click the right-arrow button to move the variable into the target Variable(s) list.

Chapter 1

▶ Click *Income category [inccat]* in the source list, and then click the right arrow button again.

Figure 1-11
Variables selected for analysis

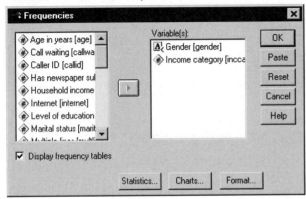

A pound sign (#) icon next to the variable name indicates that the variable is **numeric**. An icon with the letter "A" indicates that the variable is a **string** (alphanumeric) variable, which may contain both letters and numbers. A less-than sign (left angle bracket) indicates that the variable is a **short string**, containing eight or fewer characters.

▶ Click OK to run the procedure.

Viewing Results

Figure 1-12
Viewer window

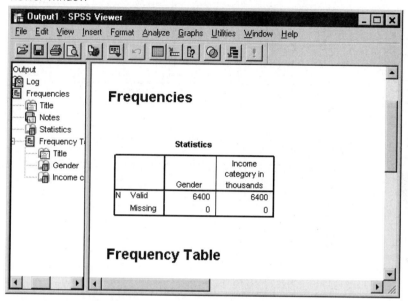

Results are displayed in the Viewer window.

You can quickly go to any item in the Viewer simply by selecting it in the outline pane.

▶ Click Income category.

Figure 1-13
Frequency table of income categories

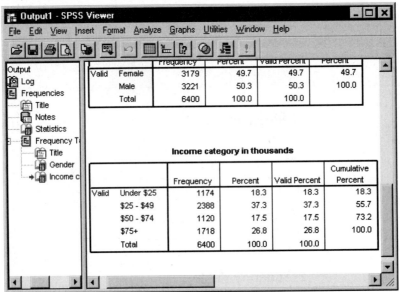

This takes you directly to the frequency table for income categories. This frequency table shows the number and percentage of people in each income category.

Creating Charts

Although some statistical procedures can create high-resolution charts, you can also use the Graphs menu to create charts.

For example, you could create a chart that shows the relationship between wireless telephone service and PDA (personal digital assistant) ownership.

▶ From the menus choose:
Graphs
   Bar...

▶ Click Clustered and then click Define.

Figure 1-14
Define Clustered Bar dialog box

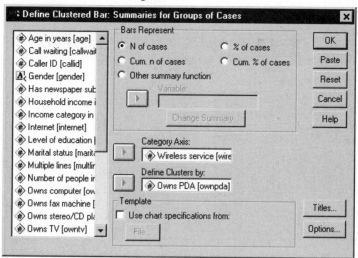

- Scroll down the source variable list and select *Wireless service [wireless]* as the Category Axis variable.

- Select *Owns PDA [ownpda]* as the Define Clusters By variable.

- Click OK to create the chart.

Chapter 1

Figure 1-15
Bar chart displayed in Viewer window

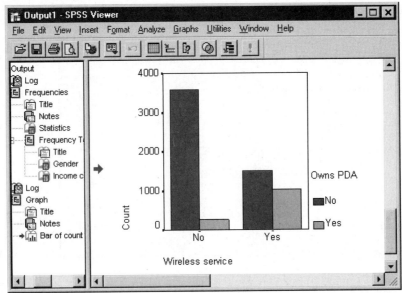

The bar chart is displayed in the Viewer. It shows that people with wireless phone service are far more likely to have PDAs than people without wireless service.

You can edit charts and tables by double-clicking on them in the contents pane of the Viewer window, and you can copy and paste your results into other applications. Those topics will be covered later.

Exiting SPSS

To exit SPSS:

▶ From the menus choose:
File
 Exit

▶ Click No if you get an alert asking if you want to save your results.

Chapter 2

Using the Help System

Help is available in a number of different ways, including:

Help menu. Every window has a Help menu on the menu bar. The Topics menu item provides access to the Help system, where you can use the Contents and Index tabs to find topics. The Tutorial menu item provides access to the introductory tutorial.

Dialog box Help buttons. Most dialog boxes have a Help button that takes you directly to a Help topic for that dialog box. The Help topic provides general information and links to related topics.

Pivot table context menu Help. Right-click on terms in an activated pivot table in the Viewer and select What's This? from the context menu to display definitions of the terms.

Statistics Coach. The Statistics Coach item on the Help menu provides a wizard-like method for finding the right statistical or charting procedure for what you want to do.

Case Studies. The Case Studies item on the Help menu provides hands-on examples of how to create various types of statistical analyses and interpret the results. The sample data files used in the examples are also provided so that you can work through the examples to see exactly how the results were produced.

This chapter uses the files *demo.sav* and *bhelptut.spo*.

Chapter 2

Help Contents Tab

The Topics item on the Help menu opens a Help window.

▶ From the menus choose:
Help
  Topics

Figure 2-1
Help Contents tab

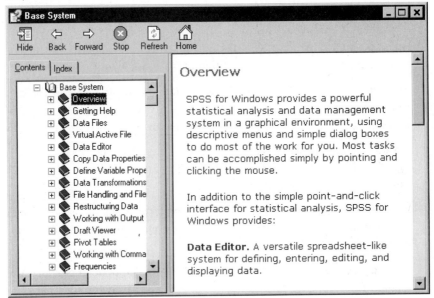

The Contents tab in the left pane of the Help window is an expandable and collapsible table of contents. It is most useful if you're looking for general information or are unsure of what index term to use to find what you're looking for.

Using the Help System

Help Index Tab

Figure 2-2
Help Index tab

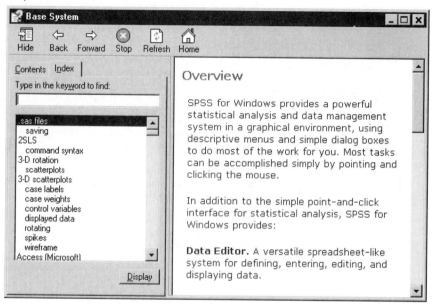

▶ Click the Index tab in the left pane of the Help window.

The Index tab provides a searchable index that makes it easy to find specific topics. The Index tab is organized in alphabetical order, just like a book index. It uses **incremental search** to find what you're looking for.

For example, you can:

▶ Type med.

Figure 2-3
Incremental index search

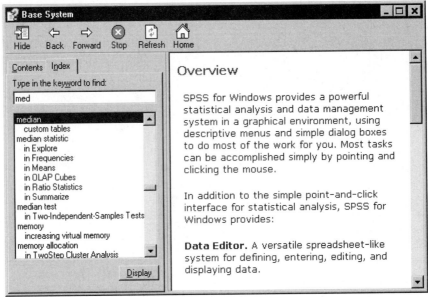

The index scrolls to and highlights the first index entry that starts with these letters, which is median.

Dialog Box Help

Most dialog boxes have a Help button that displays a Help topic about what the dialog box does and how to do it.

▶ From the menus choose:
   Analyze
     Descriptive Statistics
       Frequencies...

▶ Click Help.

Figure 2-4
Dialog box Help topic

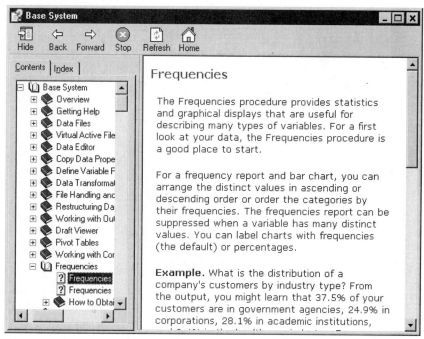

In this example, the Help topic describes the purpose of the Frequencies procedure and provides an example.

Statistics Coach

The Statistics Coach can help to guide you through the process of finding the procedure that you want to use.

▶ From the menus choose:
Help
  Statistics Coach

Figure 2-5
Statistics Coach, first step

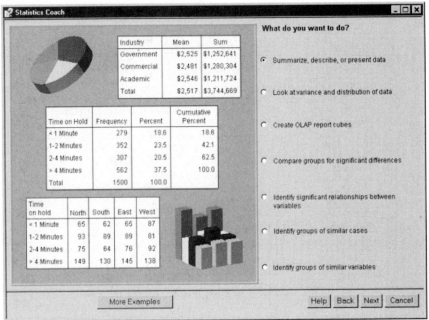

The Statistics Coach presents a series of questions designed to find the appropriate procedure. The first question is simply "What do you want to do?"

For example, if you want to summarize data:

► Select Summarize, describe, or present data.

► Then click Next.

Using the Help System

Figure 2-6
Data type selection

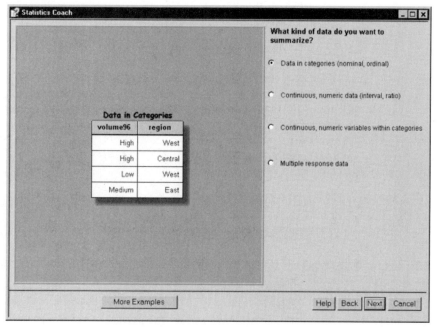

The next question asks about the type of data you want to summarize. If you're unsure, each choice displays different examples.

▶ Select Continuous, numeric data (interval, ratio).

Chapter 2

Figure 2-7
Selecting a different data type

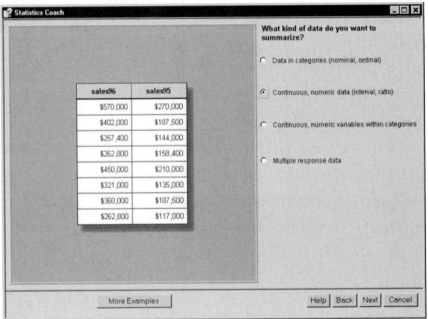

The example changes to reflect your choice. If you're still unsure, you can:

▶ Click More Examples.

A new example of the same data type is displayed. If the examples don't provide enough information, you can:

▶ Click Help.

Figure 2-8
Statistics Coach Help topic

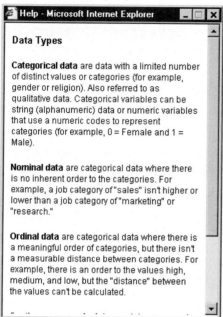

In this example, the Help topic defines the different data types.

▶ Close the Help window.

▶ Select Data in categories (nominal, ordinal) and click Next.

Chapter 2

▶ Select Tables and numbers and click Next.

Figure 2-9
Selecting tables or charts

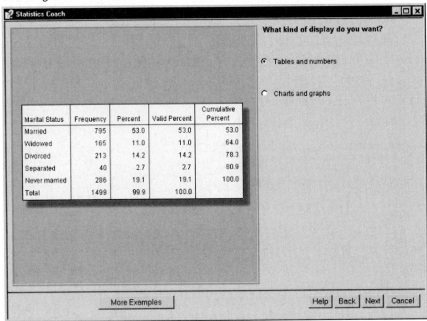

Using the Help System

▶ Select Individual case listings within categories.

Figure 2-10
Statistics Coach, final step

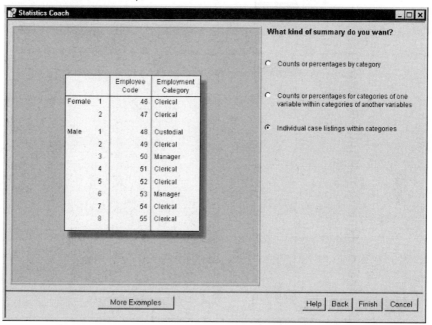

When the Statistics Coach has enough information, the Next button changes to Finish.

When you click Finish, the dialog box for the selected procedure opens automatically, and a Help topic for the procedure is also displayed.

Figure 2-11
Dialog box and Help topic

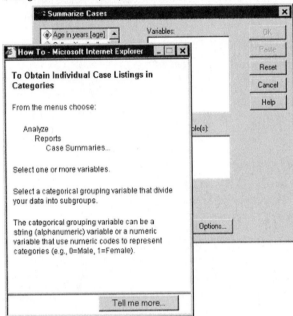

This is a custom Help topic, based on your selections in the Statistics Coach. Since some dialog boxes perform numerous functions, more than one path in the Statistics Coach may lead to the same dialog box, but the instructions in the Help topic may be different.

▶ Click Tell me more in the Help topic to get more detailed information.

Using the Help System

Figure 2-12
"Tell me more" Help topic

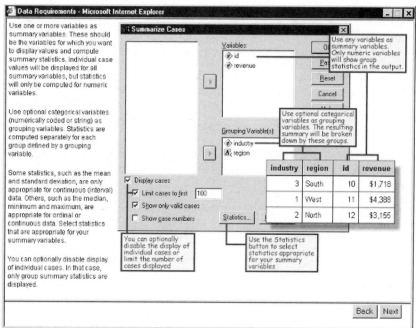

This Help topic provides detailed information on the data type(s) appropriate for the selected procedure.

Case Studies

Case studies provide comprehensive overviews of each procedure. Data files used in the examples are installed with SPSS, so you can follow along, performing the same analysis—from opening the data source and selecting variables for analysis to interpreting the results.

To access the case studies:

▶ Right-click on any pivot table created by a procedure. For example, you can right-click on the frequency table for Gender.

Figure 2-13
Accessing the case studies

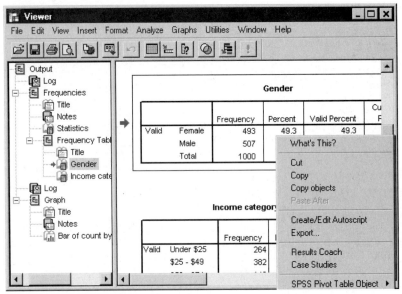

► Select Case Studies on the pop-up context menu.

Case studies are not available for all procedures. The Case Studies choice on the context menu will appear only if the feature is available for the procedure that created the selected pivot table.

Chapter 3

Reading Data

Data can be entered directly into SPSS, or it can be imported from a number of different sources. The processes for reading data stored in SPSS data files, spreadsheet applications like Microsoft Excel, database applications like Microsoft Access, and text files are all discussed in this chapter.

Basic Structure of an SPSS Data File

Figure 3-1
Data Editor

	age	marital	address	income	inccat	ca
1	37	1	12	35.00	2.00	
2	33	1	12	29.00	2.00	
3	42	1	21	34.00	2.00	
4	58	0	28	49.00	2.00	
5	56	0	9	57.00	3.00	
6	46	0	5	39.00	2.00	
7	47	1	4	56.00	3.00	
8	62	1	16	250.00	4.00	
9	52	1	27	73.00	3.00	

SPSS data files are organized by cases (rows) and variables (columns). In this data file, cases represent individual respondents to a survey. Variables represent each question asked in the survey.

Chapter 3

Reading an SPSS Data File

SPSS data files, which have a *.sav* file extension, contain your saved data. To open *demo.sav*, an example file that is installed with the product:

▶ From the menus choose:
File
 Open
  Data...

▶ Make sure SPSS (*.sav) is selected in the Files of Type drop-down list.

Figure 3-2
Open File dialog box

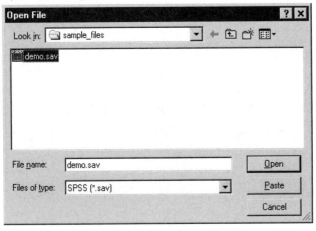

▶ Navigate to the *sample_files* folder.

▶ Select *demo.sav* and click Open.

The data are now displayed in the Data Editor.

Figure 3-3
Opened data file

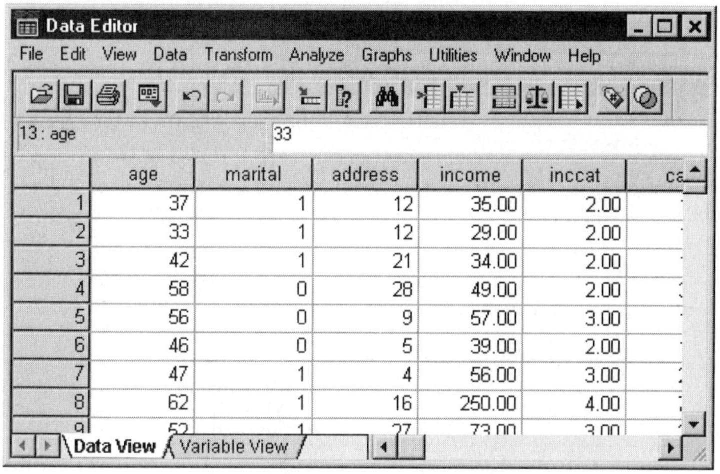

	age	marital	address	income	inccat
1	37	1	12	35.00	2.00
2	33	1	12	29.00	2.00
3	42	1	21	34.00	2.00
4	58	0	28	49.00	2.00
5	56	0	9	57.00	3.00
6	46	0	5	39.00	2.00
7	47	1	4	56.00	3.00
8	62	1	16	250.00	4.00
9	52	1	27	73.00	3.00

Reading Data from Spreadsheets

Rather than typing all of your data directly into the Data Editor, you can read data from applications like Microsoft Excel. You can also read column headings as variable names.

▶ From the menus choose:
File
  Open
    Data...

Chapter 3

▶ Select Excel (*.xls) from the Files of Type drop-down list.

Figure 3-4
Open File dialog box

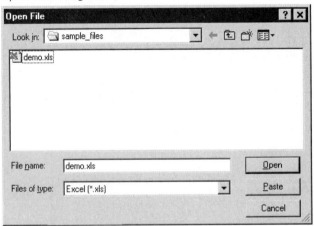

▶ Select *demo.xls* and click Open to read this spreadsheet.

The Opening Excel Data Source dialog box is displayed, allowing you to specify whether variable names are to be included in the spreadsheet, as well as the cells that you want to import. In Excel 5 or later, you can also specify which worksheets you want to import.

Figure 3-5
Opening Excel Data Source dialog box

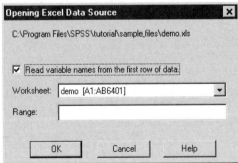

Reading Data

▶ Make sure Read variable names from first row of data is selected. This option reads column headings as variable names.

If the column headings do not conform to the SPSS variable-naming rules, they are converted into valid variable names and the original column headings are saved as variable labels. If you want to import only a portion of the spreadsheet, specify the range of cells to be imported in the Range field.

▶ Click OK to read the Excel file.

The data now appear in the Data Editor, with the column headings used as variable names. If you're using a spreadsheet application other than Excel or Lotus, you should be able to export your data to a supported format that can then be read into SPSS.

Figure 3-6
Imported Excel data

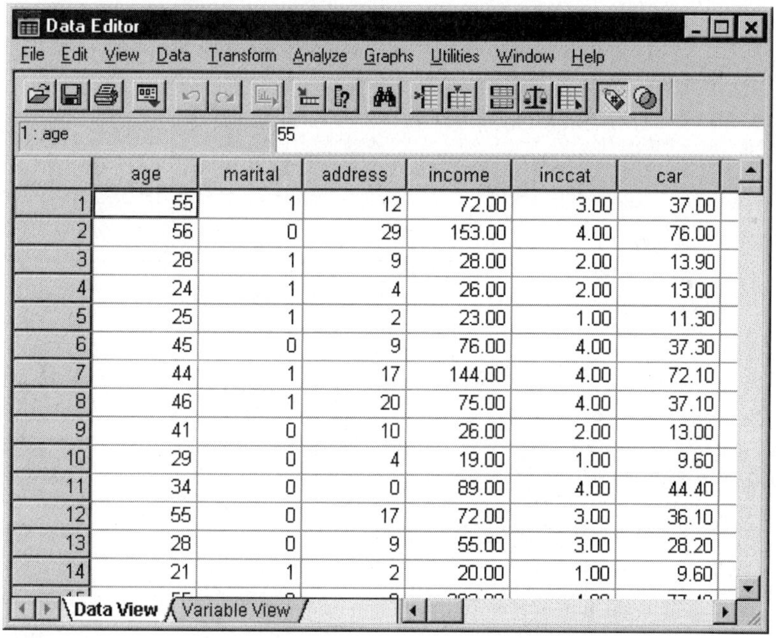

	age	marital	address	income	inccat	car
1	55	1	12	72.00	3.00	37.00
2	56	0	29	153.00	4.00	76.00
3	28	1	9	28.00	2.00	13.90
4	24	1	4	26.00	2.00	13.00
5	25	1	2	23.00	1.00	11.30
6	45	0	9	76.00	4.00	37.30
7	44	1	17	144.00	4.00	72.10
8	46	1	20	75.00	4.00	37.10
9	41	0	10	26.00	2.00	13.00
10	29	0	4	19.00	1.00	9.60
11	34	0	0	89.00	4.00	44.40
12	55	0	17	72.00	3.00	36.10
13	28	0	9	55.00	3.00	28.20
14	21	1	2	20.00	1.00	9.60

Chapter 3

Reading Data from a Database

Data from database sources are easily imported using the Database Wizard. Any database that uses ODBC (Open Database Connectivity) drivers can be read directly by SPSS after the drivers are installed. ODBC drivers for many database formats are supplied on the installation CD. Additional drivers can be obtained from third-party vendors. One of the most common database applications, Microsoft Access, is discussed in this example.

▶ From the menus choose:
File
  Open Database
    New Query...

Figure 3-7
Database Wizard Welcome dialog box

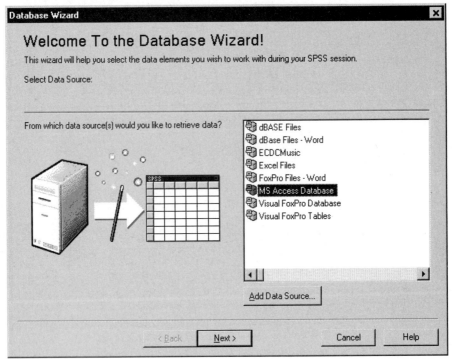

▶ Select MS Access Database from the list of data sources and click Next.

Reading Data

If MS Access Database is not listed here, you need to run *Microsoft Data Access Pack.exe*, which can be found in the Microsoft Data Access Pack folder on the CD.

Figure 3-8
ODBC Driver Login dialog box

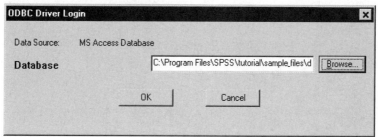

▶ Click Browse to navigate to the Access database file you want to open.

Figure 3-9
Open File dialog box

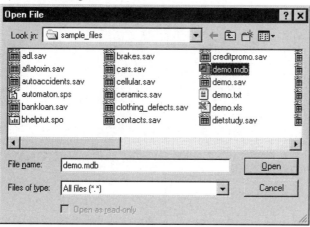

▶ Select *demo.mdb* and click Open to continue.

▶ Click OK in the login dialog box.

Chapter 3

In Step 2, you can specify the tables and variables you want to import.

Figure 3-10
Select Data dialog box

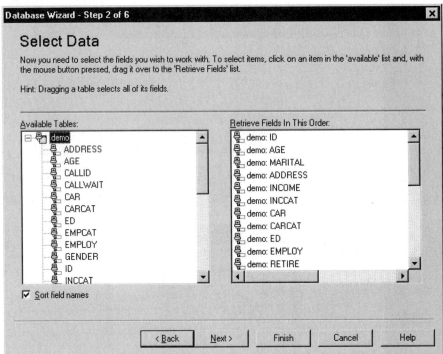

▶ Drag the entire demo table to the Retrieve Fields in This Order list.

▶ Click Next.

In Step 4, you select which records (cases) to import.

Figure 3-11
Limit Retrieved Cases dialog box

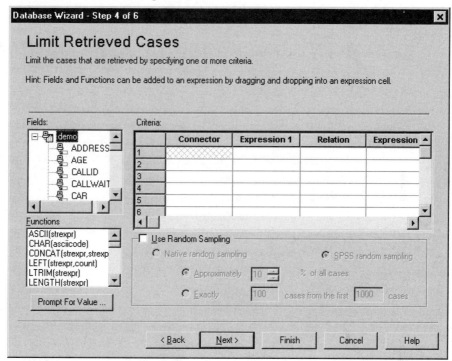

If you do not want to import all cases, you can import a subset of cases (for example, males older than 30), or you can import a random sample of cases from the data source. For large data sources, you may want to limit the number of cases to a small, representative sample to reduce the processing time. The default is to retrieve all cases.

▶ Click Next to continue.

Field names are used to create variable names. If necessary, the names are converted to valid variable names. The original field names are preserved as variable labels. You can also change the variable names before importing the database.

Figure 3-12
Define Variables dialog box

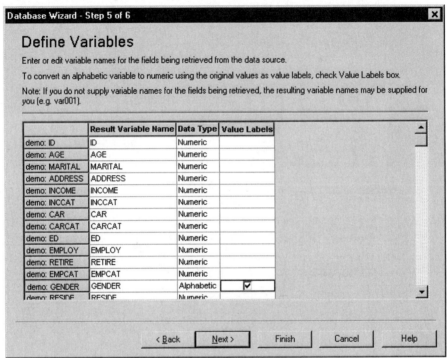

- ▶ Click the Value Labels cell in the Gender field. This option converts string variables to integer variables and retains the original value as the value label for the new variable.

- ▶ Click Next to continue.

Reading Data

The SQL statement created from your selections in the Database Wizard appears in the Results dialog box. This statement can be executed now or saved to a file for later use.

Figure 3-13
Results dialog box

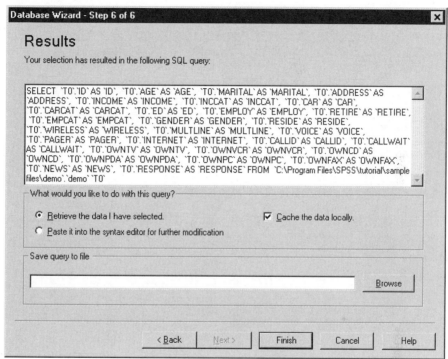

▶ Click Finish to import the data.

All of the data in the Access database that you selected to import are now available in the SPSS Data Editor.

Figure 3-14
Data imported from an Access database

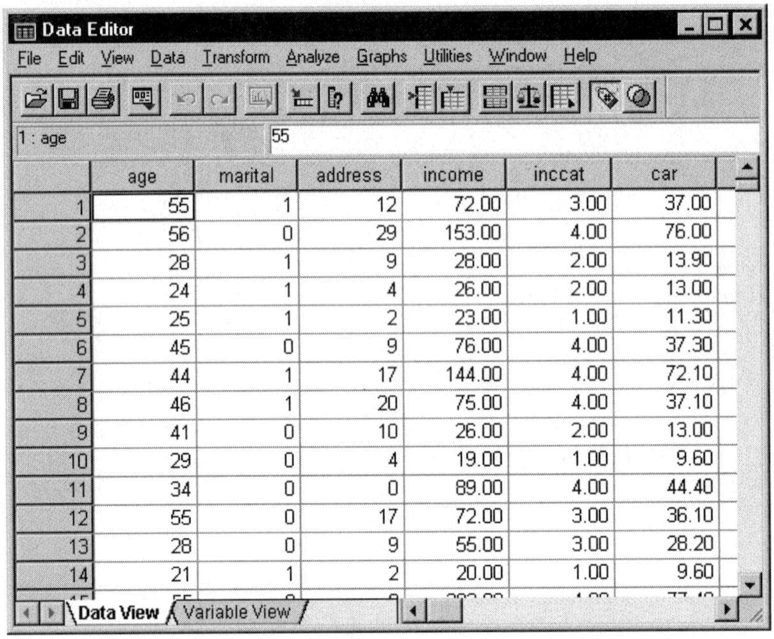
Data Editor
File Edit View Data Transform Analyze Graphs Utilities Window Help
1 : age
55
age marital address income inccat car
1 55 1 12 72.00 3.00 37.00
2 56 0 29 153.00 4.00 76.00
3 28 1 9 28.00 2.00 13.90
4 24 1 4 26.00 2.00 13.00
5 25 1 2 23.00 1.00 11.30
6 45 0 9 76.00 4.00 37.30
7 44 1 17 144.00 4.00 72.10
8 46 1 20 75.00 4.00 37.10
9 41 0 10 26.00 2.00 13.00
10 29 0 4 19.00 1.00 9.60
11 34 0 0 89.00 4.00 44.40
12 55 0 17 72.00 3.00 36.10
13 28 0 9 55.00 3.00 28.20
14 21 1 2 20.00 1.00 9.60
Data View
Variable View

Reading Data from a Text File

Text files are another common source of data. Many spreadsheet programs and databases can save their contents in one of many text file formats. Comma or tab-delimited files refer to rows of data that use commas or tabs to indicate each variable. In this example, the data are tab delimited.

▶ From the menus choose:

File
  Read Text Data...

▶ Choose Text (*.txt) from the Files of Type list.

Figure 3-15
Open File dialog box

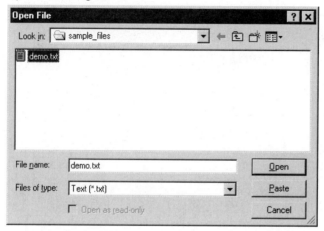

▶ Select *demo.txt* and click Open to read the selected file.

The Text Import Wizard guides you through the process of defining how the specified text file should be interpreted.

Figure 3-16
Text Import Wizard - Step 1 of 6

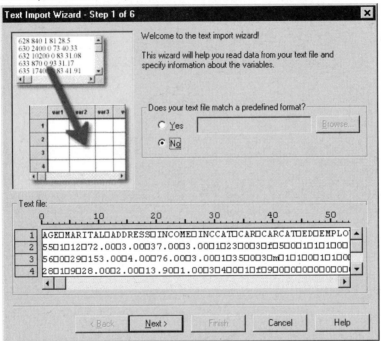

▶ In Step 1, you can choose a predefined format or create a new format in the wizard. Select No to indicate that a new format should be created.

▶ Click Next to continue.

Reading Data

As stated earlier, this file uses tab-delimited formatting. Also, the variable names are defined on the top line of this file.

Figure 3-17
Text Import Wizard - Step 2 of 6

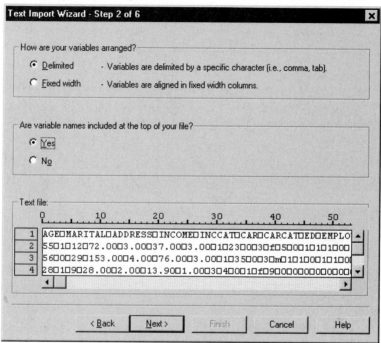

- ▶ Select Delimited to indicate that the data uses a delimited formatting structure.
- ▶ Select Yes to indicate that variable names should be read from the top of the file.
- ▶ Click Next to continue.

Chapter 3

▶ Type 2 in the top section of next dialog box to indicate that the first row of data starts on the second line of the text file.

Figure 3-18
Text Import Wizard - Step 3 of 6

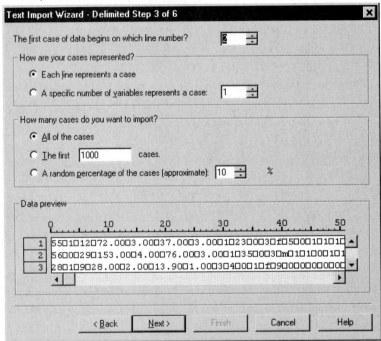

▶ Keep the default values for the remainder of this dialog box and click Next to continue.

Reading Data

The Data preview in Step 4 provides you with a quick way to ensure that your data are being properly read by SPSS.

Figure 3-19
Text Import Wizard - Step 4 of 6

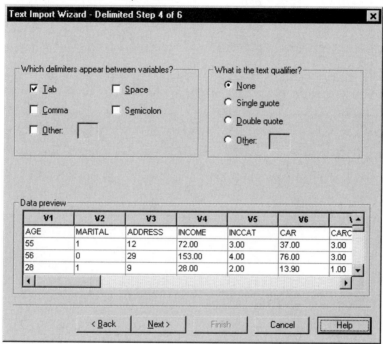

- Select Tab and deselect the other options.
- Click Next to continue.

Because the variable names may have been truncated to fit SPSS formatting requirements, this dialog box gives you the opportunity to edit any undesirable names.

Figure 3-20
Text Import Wizard - Step 5 of 6

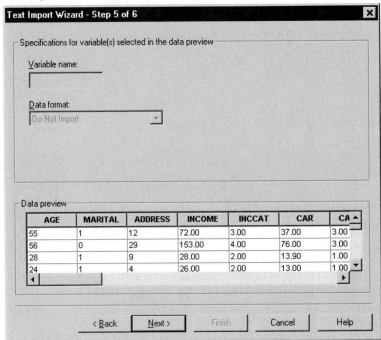

Data types can be defined here as well. For example, it's safe to assume that the income variable is meant to contain a certain dollar amount.

To change a data type:

▶ Under Data preview, select the variable you want to change, which is *Income* in this case.

▶ Select Dollar from the Data format drop-down list.

Figure 3-21
Change the data type

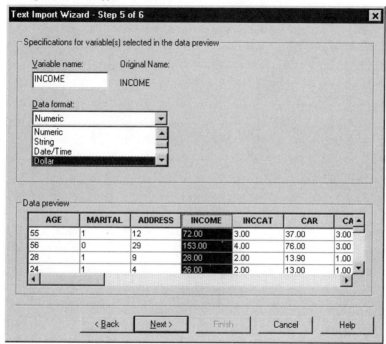

▶ Click Next to continue.

Figure 3-22
Text Import Wizard - Step 6 of 6

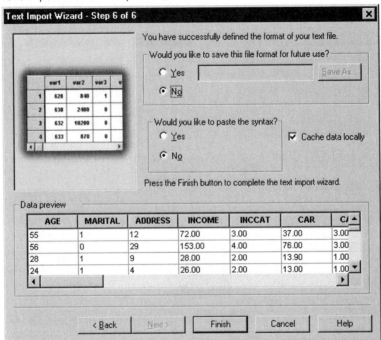

▶ Leave the default selections in this dialog box, and click Finish to import the data.

Saving Data

To save an SPSS data file, the Data Editor window must be the active window.

Reading Data

▶ From the menus choose:
File
   Save

▶ Browse to the desired directory.

▶ Type a name for the file in the File name text box.

The Variables button can be used to select which variables in the Data Editor are saved to the SPSS data file. By default, all variables in the Data Editor are retained.

▶ Click Save.

The name in the title bar of the Data Editor will change to the filename you specified. This confirms that the file has been successfully saved as an SPSS data file. The file contains both variable information (names, type, and, if provided, labels and missing value codes), and all data values.

Chapter

4

Using the Data Editor

The Data Editor displays the contents of the active data file. The information in the Data Editor consists of variables and cases.

- In Data View, columns represent variables and rows represent cases (observations).
- In Variable View, each row is a variable, and each column is an attribute associated with that variable.

Variables are used to represent the different types of data that you have compiled. A common analogy is that of a survey. The response to each question on a survey is equivalent to a variable. Variables come in many different types, including numbers, strings, currency, and dates.

Entering Numeric Data

Data can be entered into the Data Editor, which may be useful for small data files or for making minor edits to larger data files.

▶ Click the Variable View tab at the bottom of the Data Editor window.

Define the variables that are going to be used. In this case, only three variables are needed: *age*, *marital status*, and *income*.

Figure 4-1
Variable names in Variable View

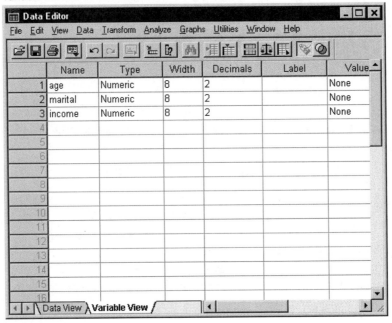

- In the first row of the first column, type age.

- In the second row, type marital.

- In the third row, type income.

  New variables are automatically given a numeric data type.

  If you don't enter variable names, unique names are automatically created. However, these names are not descriptive and are not recommended for large data files.

- Click the Data View tab to continue entering the data.

  The names that you entered in Variable View are now the headings for the first three columns in Data View.

Begin entering data in the first row, starting at the first column.

Figure 4-2
Values entered in Data View

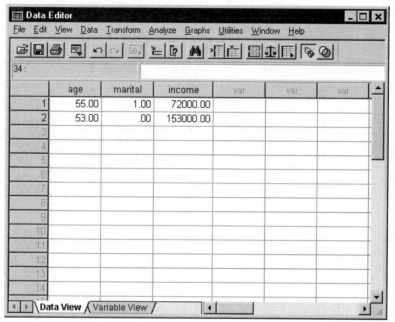

- In the *age* column, type 55.
- In the *marital* column, type 1.
- In the *income* column, type 72000.
- Move the cursor to the first column of the second row to add the next subject's data.
- In the *age* column, type 53.
- In the *marital* column, type 0.
- In the *income* column, type 153000.

Currently, the *age* and *marital* columns display decimal points, even though their values are intended to be integers. To hide the decimal points in these variables:

▶ Click the Variable View tab at the bottom of the Data Editor window.

▶ Select the *Decimals* column in the *age* row and type 0 to hide the decimal.

▶ Select the *Decimals* column in the *marital* row and type 0 to hide the decimal.

Figure 4-3
Updated decimal property for age and marital

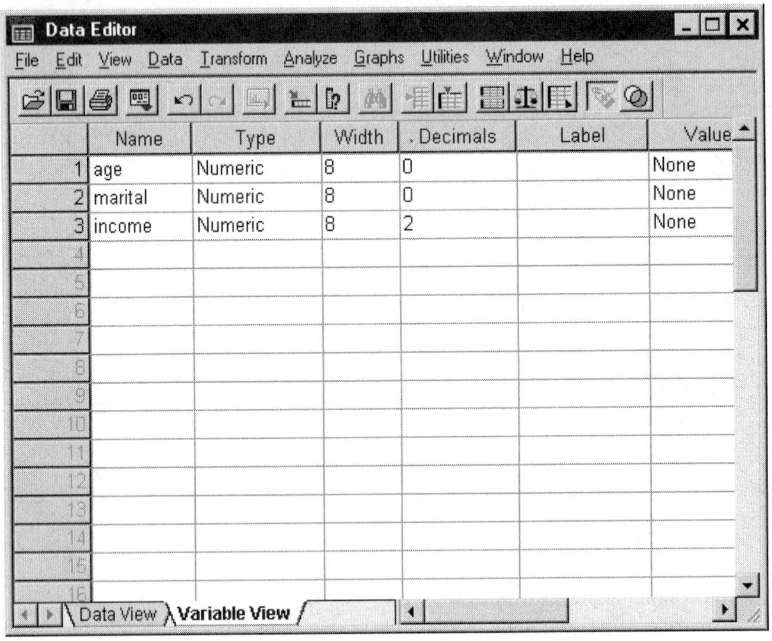

	Name	Type	Width	Decimals	Label	Value
1	age	Numeric	8	0		None
2	marital	Numeric	8	0		None
3	income	Numeric	8	2		None

Entering String Data

Non-numeric data, such as strings of text, can also be entered into the Data Editor.

▶ Click the Variable View tab at the bottom of the Data Editor window.

▶ In the first cell of the first empty row, type **sex** for the variable name.

Using the Data Editor

▶ Click the *Type* cell.

▶ Click the button in the *Type* cell to open the Variable Type dialog box.

Figure 4-4
Button shown in Type cell for sex

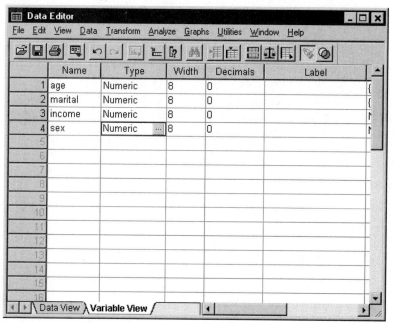

▶ Select String to specify the variable type.

Chapter 4

▶ Click OK to save your changes and return to the Data Editor.

Figure 4-5
Variable Type dialog box

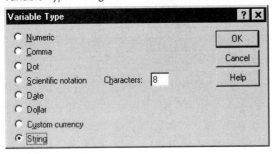

Defining Data

In addition to defining data types, you can also define descriptive variable and value labels for variable names and data values. These descriptive labels are used in statistical reports and charts.

Adding a Variable Label

Labels are meant to provide descriptions of variables. These descriptions are often longer versions of variable names. Labels can be up to 256 characters long. These labels are used in your output to identify the different variables.

▶ Click the Variable View tab at the bottom of the Data Editor window.

▶ In the *Label* column of the *age* row, type Respondent's Age.

▶ In the *Label* column of the *marital* row, type Marital Status.

▶ In the *Label* column of the *income* row, type Household Income.

Using the Data Editor

▶ In the *Label* column of the *sex* row, type **Gender**.

Figure 4-6
Variable labels entered in Variable View

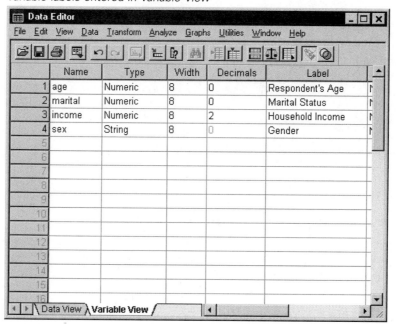

Changing Variable Type and Format

The *Type* column displays the current data type for each variable. The most common are numeric and string, but many other formats are supported. In the current data file, the *income* variable is defined as a numeric type.

▶ Click the *Type* cell for the *income* row, and then click the button to open the Variable Type dialog box.

▶ Select Dollar in the Variable Type dialog box.

Figure 4-7
Variable Type dialog box

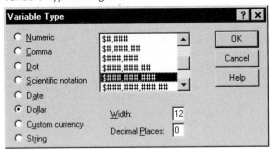

The formatting options for the currently selected data type are displayed.

▶ Select the format of this currency. For this example, select $###,###,###.

▶ Click OK to save your changes.

Adding Value Labels for Numeric Variables

Value labels provide a method for mapping your variable values to a string label. In the case of this example, there are two acceptable values for the *marital* variable. A value of 0 means that the subject is single, and a value of 1 means that he or she is married.

▶ Click the *Values* cell for the *marital* row, and then click the button to open the Value Labels dialog box.

The **value** is the actual numeric value.

The **value label** is the string label applied to the specified numeric value.

▶ Type 0 in the Value field.

▶ Type Single in the Value Label field.

Using the Data Editor

▶ Click Add to add this label to the list.

Figure 4-8
Value Labels dialog box

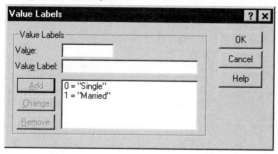

▶ Repeat the process, this time typing 1 in the Value field and Married in the Value Label field.

▶ Click Add, and then click OK to save your changes and return to the Data Editor.

These labels can also be displayed in Data View, which can help to make your data more readable.

▶ Click the Data View tab at the bottom of the Data Editor window.

▶ From the menus choose:
View
  Value Labels

The labels are now displayed in a list when you enter values in the Data Editor. This has the benefit of suggesting a valid response and providing a more descriptive answer.

Figure 4-9
Value labels displayed in Data View

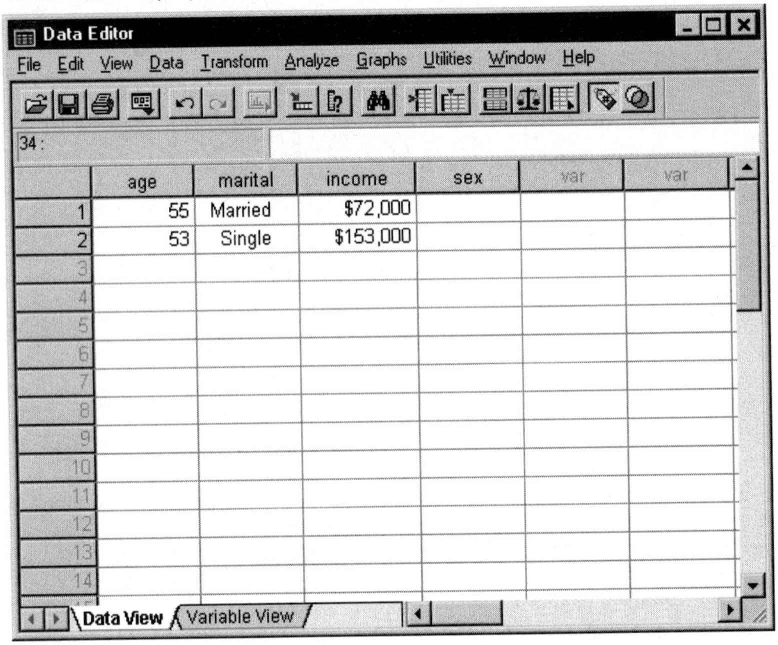

	age	marital	income	sex
1	55	Married	$72,000	
2	53	Single	$153,000	

Adding Value Labels for String Variables

String variables may require value labels as well. For example, your data may use single letters, *M* or *F*, to identify the sex of the subject. Value labels can be used to specify that *M* stands for *Male* and *F* stands for *Female*.

▶ Click the Variable View tab at the bottom of the Data Editor window.

▶ Click the *Values* cell in the *sex* row, and then click the button to open the Value Labels dialog box.

▶ Type F in the Value field, and then type Female in the Value Label field.

Using the Data Editor

▶ Click Add to add this label to your data file.

Figure 4-10
Value Labels dialog box

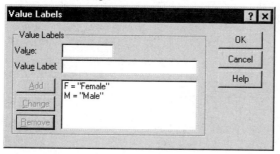

▶ Repeat the process, this time typing M in the Value field and Male in the Value Label field.

▶ Click Add, and then click OK to save your changes and return to the Data Editor.

Because string values are case sensitive, you should make sure that you are consistent. A lowercase *m* is not the same as an uppercase *M*.

Using Value Labels for Data Entry

In a previous example, we chose to have value labels displayed rather than the actual data by selecting Value Labels from the View menu. You can use these values for data entry.

▶ Click the Data View tab at the bottom of the Data Editor window.

▶ In the first row, select the cell for *sex* and select Male from the drop-down list.

▶ In the second row, select the cell for *sex* and select Female from the drop-down list.

Figure 4-11
Using variable labels to select values

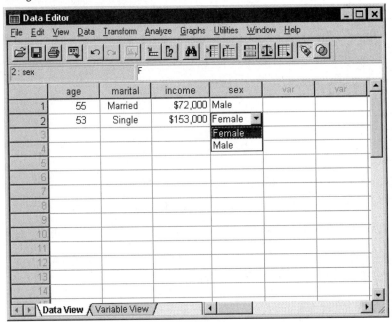

Only defined values are listed, which helps to ensure that the data entered are in a format that you expect.

Handling Missing Data

Missing or invalid data are generally too common to ignore. Survey respondents may refuse to answer certain questions, may not know the answer, or may answer in an unexpected format. If you don't take steps to filter or identify these data, your analysis may not provide accurate results.

For numeric data, empty data fields or fields containing invalid entries are handled by converting the fields to system missing, which is identifiable by a single period.

Using the Data Editor

Figure 4-12
Missing values displayed as periods

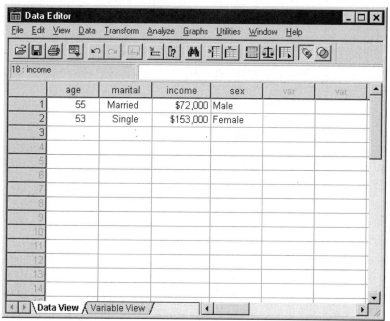

The reason a value is missing may be important to your analysis. For example, you may find it useful to distinguish between those who refused to answer a question and those who didn't answer a question because it was not applicable.

Missing Values for a Numeric Variable

▶ Click the Variable View tab at the bottom of the Data Editor window.

▶ Click the *Missing* cell in the *age* row, and then click the button to open the Missing Values dialog box.

Chapter 4

In this dialog box, you can specify up to three distinct missing values, or a range of values plus one additional discrete value.

Figure 4-13
Missing Values dialog box

▶ Select Discrete missing values.

▶ Type 999 in the first text box and leave the other two empty.

▶ Click OK to save your changes and return to the Data Editor.

Now that the missing data value has been added, a label can be applied to that value.

▶ Click the *Values* cell in the *age* row, and then click the button to open the Value Labels dialog box.

▶ Type 999 in the Value field.

▶ Type No Response in the Value Label field.

Figure 4-14
Value Labels dialog box

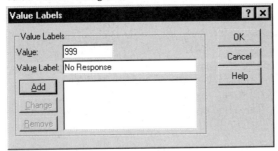

Using the Data Editor

▶ Click Add to add this label to your data file.

▶ Click OK to save your changes and return to the Data Editor.

Missing Values for a String Variable

Missing values for string variables are handled similarly to those for numeric values. Unlike numeric values, empty fields in string variables are not designated as system missing. Rather, they are interpreted as an empty string.

▶ Click the Variable View tab at the bottom of the Data Editor window.

▶ Click the *Missing* cell in the *sex* row, and then click the button to open the Missing Values dialog box.

▶ Select Discrete missing values.

▶ Type NR in the first text box.

Missing values for string variables are case sensitive. So, a value of *nr* is not treated as a missing value.

▶ Click OK to save your changes and return to the Data Editor.

Now you can add a label for the missing value.

▶ Click the *Values* cell in the *sex* row, and then click the button to open the Value Labels dialog box.

▶ Type NR in the Value field.

▶ Type No Response in the Value Label field.

Figure 4-15
Value Labels dialog box

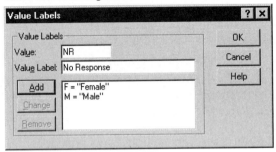

▶ Click Add to add this label to your project.

▶ Click OK to save your changes and return to the Data Editor.

Copying and Pasting Value Attributes

Once you've defined variable attributes for a variable, you can copy these attributes and apply them to other variables.

Using the Data Editor

▶ In Variable View, type **agewed** in the first cell of the first empty row.

Figure 4-16
agewed variable in Variable View

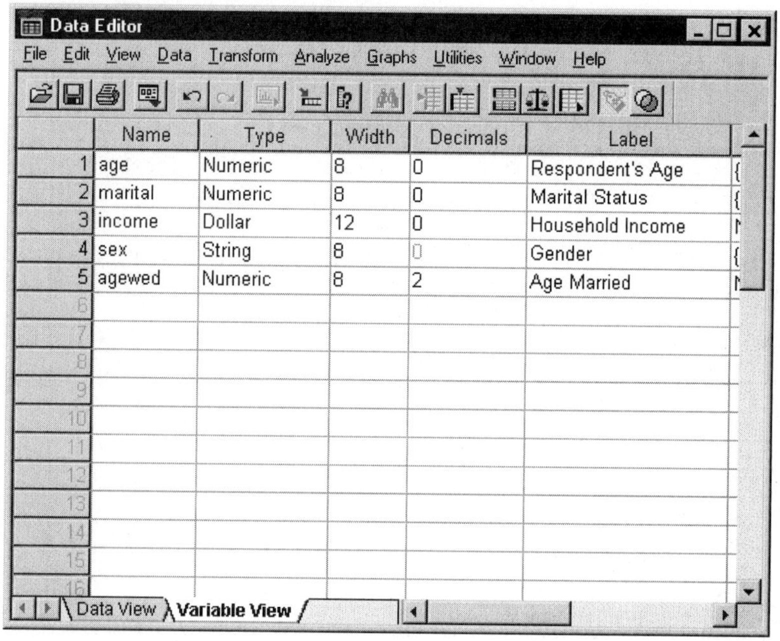

	Name	Type	Width	Decimals	Label	
1	age	Numeric	8	0	Respondent's Age	
2	marital	Numeric	8	0	Marital Status	
3	income	Dollar	12	0	Household Income	
4	sex	String	8	0	Gender	
5	agewed	Numeric	8	2	Age Married	

▶ In the *Label* column, type **Age Married**.

▶ Click the *Values* cell in the *age* row.

▶ From the menus choose:
Edit
  Copy

▶ Click the *Values* cell in the *agewed* row.

▶ From the menus choose:
Edit
  Paste

Chapter 4

The defined values from the *age* variable are now applied to the *agewed* variable.

Figure 4-17
Values pasted for agewed

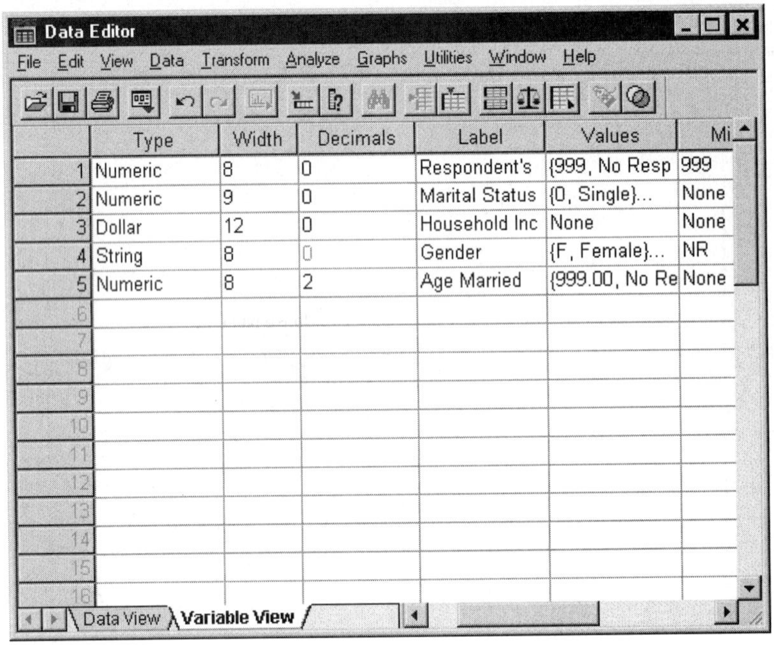

	Type	Width	Decimals	Label	Values	Mi...
1	Numeric	8	0	Respondent's	{999, No Resp	999
2	Numeric	9	0	Marital Status	{0, Single}...	None
3	Dollar	12	0	Household Inc	None	None
4	String	8	0	Gender	{F, Female}...	NR
5	Numeric	8	2	Age Married	{999.00, No Re	None

Using the Data Editor

To apply the attribute to multiple variables, simply select multiple target cells (click and drag down the column).

Figure 4-18
Multiple cells selected

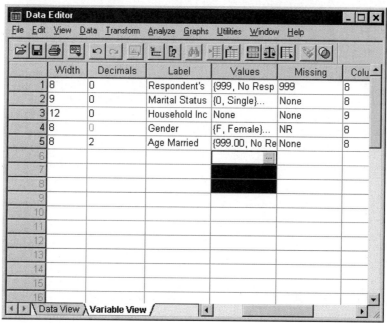

When you paste the attribute, it is applied to all of the selected cells.

Figure 4-19
Values pasted into multiple cells

	Name	Type	Width	Decimals	Label	Value
1	age	Numeric	8	0	Respondent's	{999, No
2	marital	Numeric	9	0	Marital Status	{0, Single
3	income	Dollar	12	0	Household Inc	None
4	sex	String	8	0	Gender	{F, Fema
5	agewed	Numeric	8	2	Age Married	{999.00,
6	var00001	Numeric	8	2		{999.00,
7	var00002	Numeric	8	2		{999.00,
8	var00003	Numeric	8	2		{999.00,

New variables are automatically created if you paste the values into empty rows.

You can also copy all of the attributes from one variable to another.

Using the Data Editor

▶ Click the *row number* in the *marital* row.

Figure 4-20
Selected row

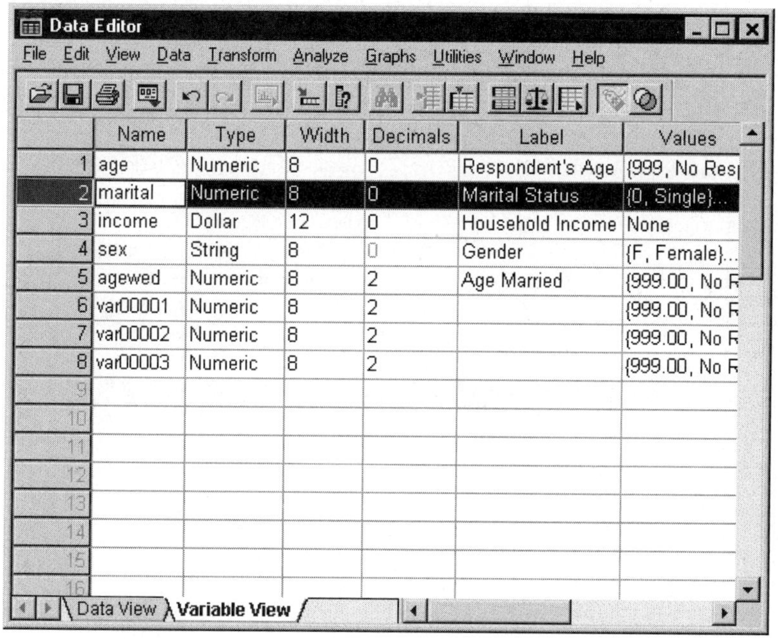

Name	Type	Width	Decimals	Label	Values
1 age	Numeric	8	0	Respondent's Age	{999, No Res
2 marital	Numeric	8	0	Marital Status	{0, Single}...
3 income	Dollar	12	0	Household Income	None
4 sex	String	8	0	Gender	{F, Female}...
5 agewed	Numeric	8	2	Age Married	{999.00, No R
6 var00001	Numeric	8	2		{999.00, No R
7 var00002	Numeric	8	2		{999.00, No R
8 var00003	Numeric	8	2		{999.00, No R

▶ From the menus choose:
Edit
 Copy

▶ Click the row number of the first empty row.

▶ From the menus choose:
Edit
 Paste

All of the attributes of the *marital* variable are applied to the new variable.

Figure 4-21
Values pasted into row

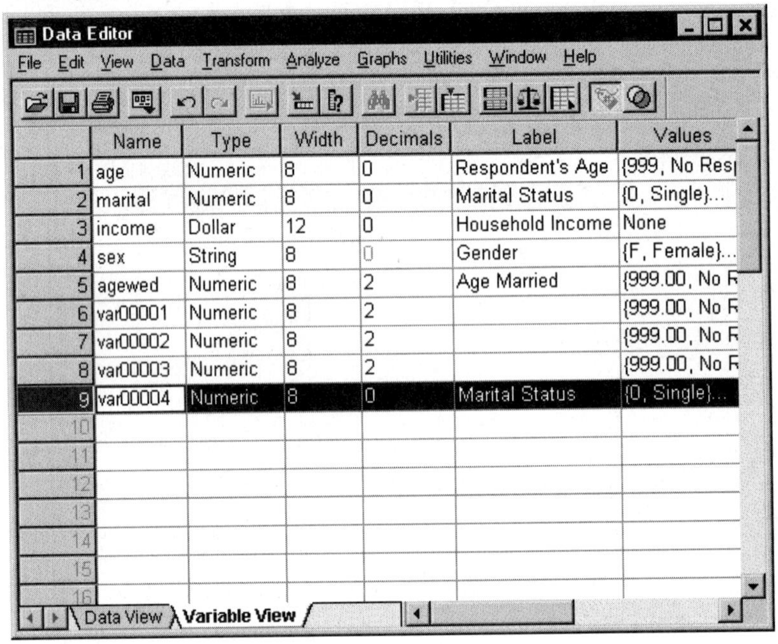

	Name	Type	Width	Decimals	Label	Values
1	age	Numeric	8	0	Respondent's Age	{999, No Resp
2	marital	Numeric	8	0	Marital Status	{0, Single}...
3	income	Dollar	12	0	Household Income	None
4	sex	String	8	0	Gender	{F, Female}...
5	agewed	Numeric	8	2	Age Married	{999.00, No R
6	var00001	Numeric	8	2		{999.00, No R
7	var00002	Numeric	8	2		{999.00, No R
8	var00003	Numeric	8	2		{999.00, No R
9	var00004	Numeric	8	0	Marital Status	{0, Single}...

Defining Variable Properties for Categorical Variables

For categorical (nominal, ordinal) data, Define Variable Properties can help you define value labels and other variable properties. Define Variable Properties:

- Scans the actual data values and lists all unique data values for each selected variable.
- Identifies unlabeled values and provides an "auto-label" feature.
- Provides the ability to copy defined value labels from another variable to the selected variable or from the selected variable to multiple additional variables.

Using the Data Editor

This example uses the data file *demo.sav*. This data file already has defined value labels; so before we start, let's enter a value for which there is no defined value label:

▶ In Data View of the Data Editor, click the first data cell for the variable *ownpc* (you may have to scroll to the right) and enter the value 99.

▶ From the menus choose:
Data
  Define Variable Properties...

Figure 4-22
Initial Define Variable Properties dialog box

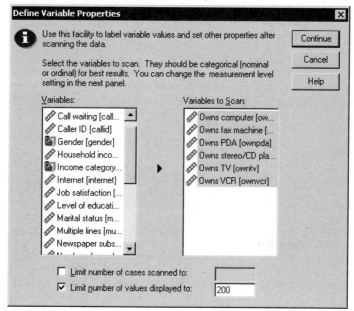

In the initial Define Variable Properties dialog box, you select the nominal or ordinal variables for which you want to define value labels and/or other properties.

Since Define Variable Properties relies on actual values in the data file to help you make good choices, it needs to read the data file first. This can take some time if your data file contains a very large number of cases, so this dialog box also allows you to limit the number of cases to read, or scan. Limiting the number of cases is not necessary for our sample data file. Even though it contains over 6,000 cases, it doesn't take very long to scan that many cases.

▶ Drag and drop *Owns computer [ownpc]* through *Owns VCR [ownvcr]* into the Variables to Scan list.

You might notice that the measurement level icons for all of the selected variables indicate that they are scale variables, not categorical variables. By default, all numeric variables are assigned the scale measurement level, even if the numeric values are actually just codes that represent categories. All of the selected variables in this example are really categorical variables that use the numeric values 0 and 1 to stand for *No* and *Yes*, respectively—and one of the variable properties that we'll change with Define Variable Properties is the measurement level.

▶ Click Continue.

Figure 4-23
Define Variable Properties main dialog box

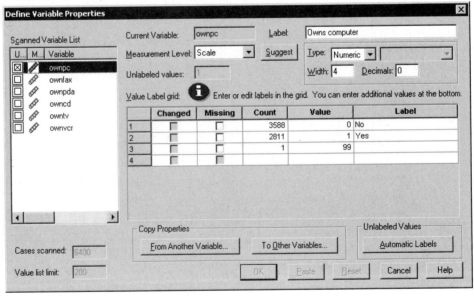

▶ In the Scanned Variable List, select *ownpc*.

Using the Data Editor

The current level of measurement for the selected variable is scale. You can change the measurement level by selecting one from the drop-down list or you can let Define Variable Properties suggest a measurement level.

▶ Click Suggest.

Figure 4-24
Suggest Measurement Level dialog box

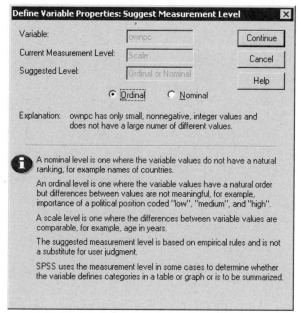

Since the variable doesn't have very many different values and all of the scanned cases contain integer values, the proper measurement level is probably ordinal or nominal.

▶ Select Ordinal and then click Continue.

The measurement level for the selected variable is now ordinal.

The Value Labels Grid displays all of the unique data values for the selected variable, any defined value labels for these values, and the number of times (count) each value occurs in the scanned cases.

The value that we entered, 99, is displayed in the grid. The count is only 1 because we changed the value for only one case, and the *Label* column is empty because we haven't defined a value label for 99 yet. An X in the first column of the Scanned

Chapter 4

Variable List also indicates that the selected variable has at least one observed value without a defined value label.

▶ In the *Label* column for the value of 99, enter **No answer**.

▶ Then click (check) the box in the *Missing* column. This identifies the value 99 as **user missing**. Data values specified as user missing are flagged for special treatment and are excluded from most calculations.

Figure 4-25
New variable properties defined for ownpc

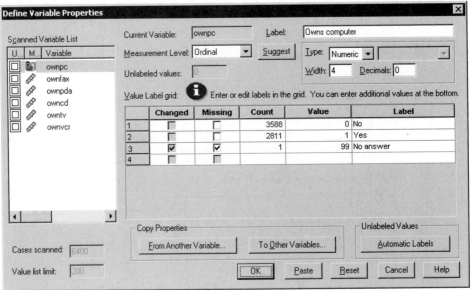

Before we complete the job of modifying the variable properties for *ownpc*, let's apply the same measurement level, value labels, and missing values definitions to the other variables in the list.

▶ In the Copy Properties group, click **To Other Variables**.

Figure 4-26
Apply Labels and Level dialog box

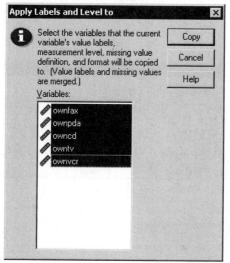

In the Apply Labels and Level dialog box, select all of the variables in the list, and then click Copy.

If you select any other variable in the list in the Define Variable Properties main dialog box now, you'll see that they are all now ordinal variables, with a value of 99 defined as user missing and a value label of *No answer*.

Figure 4-27
New variable properties defined for ownfax

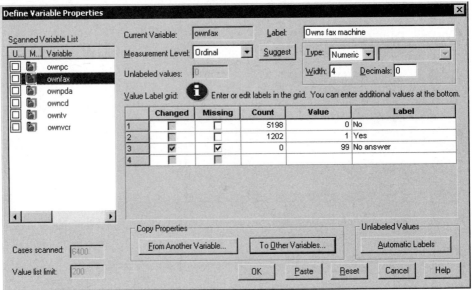

▶ Click OK to save all of the variable properties that you have defined.

Chapter

5

Examining Summary Statistics for Individual Variables

This chapter discusses simple summary measures and how the level of measurement of a variable influences the types of statistics that should be used. We will use the data file *demo.sav*.

Level of Measurement

Different summary measures are appropriate for different types of data, depending on the level of measurement:

Categorical. Data with a limited number of distinct values or categories (for example, gender or marital status). Also referred to as qualitative data. Categorical variables can be string (alphanumeric) data or numeric variables that use numeric codes to represent categories (for example, 0 = Unmarried and 1 = Married). There are two basic types of categorical data:

- **Nominal.** Categorical data where there is no inherent order to the categories. For example, a job category of "sales" isn't higher or lower than a job category of "marketing" or "research."

- **Ordinal.** Categorical data where there is a meaningful order of categories, but there isn't a measurable distance between categories. For example, there is an order to the values high, medium, and low, but the "distance" between the values can't be calculated.

Scale. Data measured on an interval or ratio scale, where the data values indicate both the order of values and the distance between values. For example, a salary of $72,195 is higher than a salary of $52,398, and the distance between the two values is $19,797. Also referred to as quantitative or continuous data.

Summary Measures for Categorical Data

For categorical data, the most typical summary measure is the number or percentage of cases in each category. The **mode** is the category with the greatest number of cases. For ordinal data, the **median** (the value above and below which half the cases fall) may also be a useful summary measure if there is a large number of categories.

The Frequencies procedure produces frequency tables that display both the number and percentage of cases for each observed value of a variable.

▶ From the menus choose:
Analyze
  Descriptive Statistics
    Frequencies...

▶ Select *Owns PDA (ownpda)* and *Owns TV (owntv)* and move them into the Variable(s) list.

Figure 5-1
Categorical variables selected for analysis

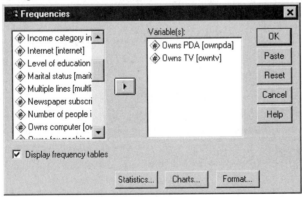

▶ Click OK to run the procedure.

Examining Summary Statistics for Individual Variables

Figure 5-2
Frequency tables

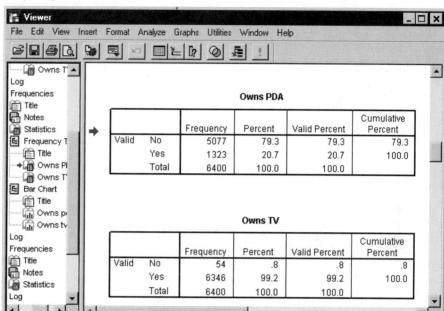

The frequency tables are displayed in the Viewer window. The frequency tables reveal that only about 21% of the people own PDAs, but almost everybody owns a TV (99.2%). This might not be an interesting revelation, although it might be interesting to find out more about the small group of people who do not own televisions.

Charts for Categorical Data

You can graphically display the information in a frequency table with a bar chart or pie chart.

▶ Open the Frequencies dialog box again. (The two variables should still be selected.)

You can use the Dialog Recall button on the toolbar to quickly return to recently used procedures.

Figure 5-3
Dialog Recall tool

▶ Click Charts.

▶ Select Bar charts and then click Continue.

Figure 5-4
Frequencies Charts dialog box

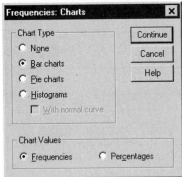

▶ Click OK in the main dialog box to run the procedure.

Examining Summary Statistics for Individual Variables

Figure 5-5
Bar chart

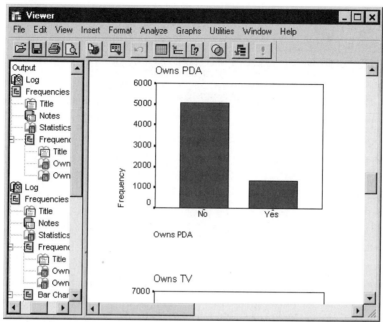

In addition to the frequency tables, the same information is now displayed in the form of bar charts, making it easy to see that most people do not own PDAs but almost everyone owns a TV.

Summary Measures for Scale Variables

There are many summary measures available for scale variables, including:

- **Measures of central tendency.** The most common measures of central tendency are the **mean** (arithmetic average) and **median** (value above and below which half the cases fall).

- **Measures of dispersion.** Statistics that measure the amount of variation or spread in the data include the standard deviation, minimum, and maximum.

▶ Open the Frequencies dialog box again.

Chapter 5

▶ Click Reset to clear any previous settings.

▶ Select *Household income in thousands (income)* and move it into the Variable(s) list.

Figure 5-6
Scale variable selected for analysis

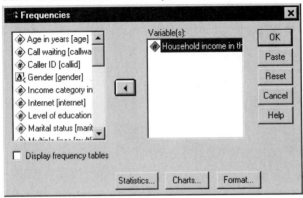

▶ Click Statistics.

▶ Select Mean, Median, Std. deviation, Minimum, and Maximum.

Figure 5-7
Frequencies Statistics dialog box

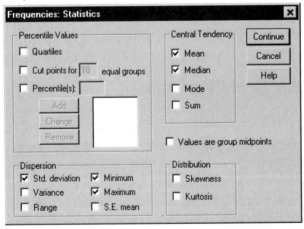

▶ Click Continue.

Examining Summary Statistics for Individual Variables

▶ Deselect Display frequency tables in the main dialog box. (Frequency tables are usually not useful for scale variables since there may be almost as many distinct values as there are cases in the data file.)

▶ Click OK to run the procedure.

The Frequencies Statistics table is displayed in the Viewer window.

Figure 5-8
Frequencies Statistics table

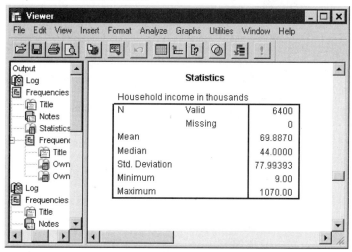

In this example, there is a large difference between the mean and the median, with the mean being more than 25,000 greater than the median. This indicates that the values are not normally distributed. We can visually check the distribution with a histogram.

Chapter 5

Histograms for Scale Variables

▶ Open the Frequencies dialog box again.

▶ Click Charts.

▶ Select Histograms and With normal curve.

Figure 5-9
Frequencies Charts dialog box

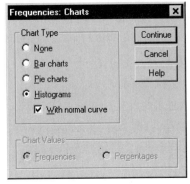

▶ Click Continue and click OK in the main dialog box to run the procedure.

Figure 5-10
Histogram

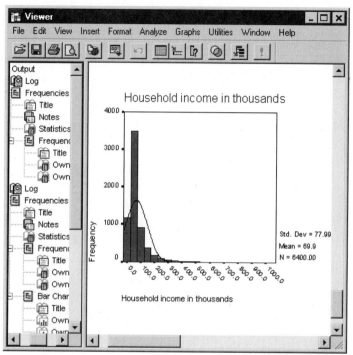

The majority of cases are clustered at the lower end of the scale, with most falling below 100,000. There are, however, a few cases in the 500,000 range and beyond (too few to even be visible without modifying the histogram). These high values for only a few cases have a significant effect on the mean but little or no effect on the median, making the median a better indicator of central tendency in this example.

Chapter 6

Working with Output

The results from running a statistical procedure are displayed in the Viewer. The output produced can be statistical tables, charts or graphs, or text, depending on the choices you make when you run the procedure. This chapter uses the files *viewertut.spo* and *demo.sav*.

Using the Viewer

Figure 6-1
SPSS Viewer

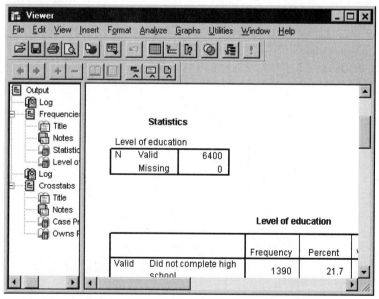

Chapter 6

The Viewer window is divided into two panes. The **outline pane** contains an outline of all of the information stored in the Viewer. The **contents pane** contains statistical tables, charts, and text output.

Use the scrollbars to navigate through the window's contents, both vertically and horizontally. For easier navigation, click an item in the outline pane to display it in the contents pane.

If you find that there isn't enough room in the Viewer to see an entire table or that the outline view is too narrow, you can easily resize the window.

▶ Click and drag the right border of the outline pane to change its width.

An open book icon in the outline pane indicates that it is currently visible in the Viewer, although it may not currently be in the visible portion of the contents pane.

▶ To hide a table or chart, double-click its book icon in the outline pane.

The open book icon changes to a closed book icon, signifying that the information associated with it is now hidden.

▶ To redisplay the hidden output, double-click the closed book icon.

You can also hide all of the output from a particular statistical procedure or all of the output in the Viewer.

Working with Output

▶ Click the box with the minus sign (-) to the left of the procedure whose results you want to hide, or click the box next to the topmost item in the outline pane to hide all of the output.

Figure 6-2
Hidden output in the Viewer

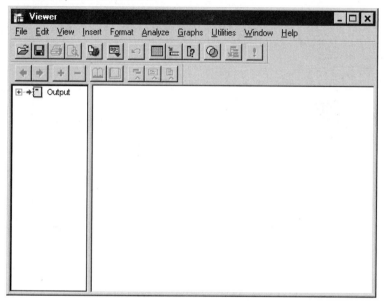

The outline collapses, visually indicating that these results are hidden.

You can also change the order in which the output is displayed.

▶ In the outline pane, click on the items you want to move.

Chapter 6

▶ Drag the selected items to a new location in the outline and release the mouse button.

Figure 6-3
Reordered output in the Viewer

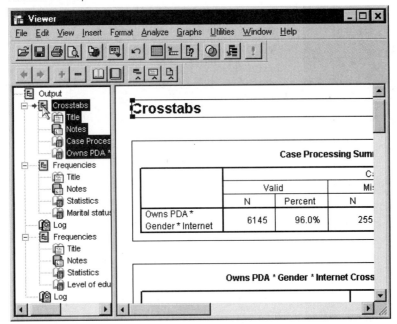

You can also move output items by clicking and dragging them in the contents pane.

Using the Pivot Table Editor

The results from most statistical procedures are displayed in **pivot tables**.

Accessing Output Definitions

Many statistical terms are displayed in the output. Definitions of these terms can be accessed directly in the Viewer.

▶ Double-click the *Owns PDA * Gender * Internet Crosstabulation* table.

Working with Output

▶ Right-click *Expected Count* and select What's This? from the pop-up context menu.

The definition is displayed in a pop-up window.

Figure 6-4
Pop-up definition

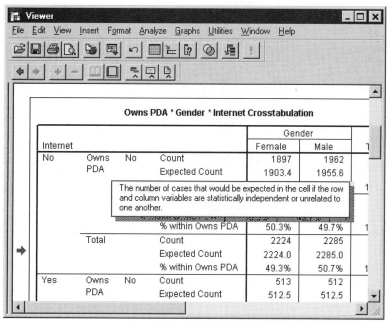

Pivoting Tables

The default tables produced may not display information as neatly or as clearly as you would like. With pivot tables, you can transpose rows and columns ("flip" the table), adjust the order of data in a table, and modify the table in many other ways. For example, you can change a short, wide table into a long, thin one by transposing rows and columns. Changing the layout of the table does not affect the results. Instead, it's a way to display your information in a different or more desirable manner.

▶ Double-click the *Owns PDA * Gender * Internet Crosstabulation* table.

Chapter 6

▶ If the Pivoting Trays window is not visible, from the menus choose:
Pivot
  Pivoting Trays

Pivoting Trays provide a way to move data between columns, rows, and layers.

Figure 6-5
Pivoting Trays

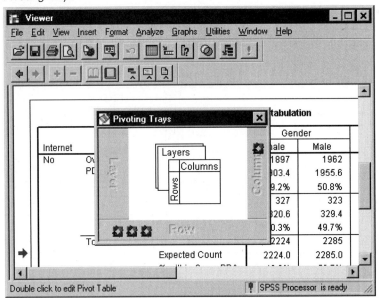

Working with Output

▶ Click one of the pivot icons to see what it represents. The shaded area in the table indicates what will be moved when you move the pivot icon. A pop-up label also indicates what the icon represents in the table.

Figure 6-6
Pivot icon labels

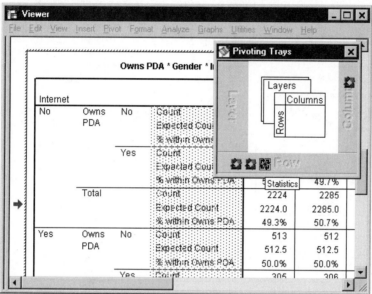

▶ Drag the Statistics pivot icon from the Row dimension to the bottom of the Column dimension. The table is immediately reconfigured to reflect your changes.

The order of the pivot icons in a dimension reflects the order of the elements in the table.

Figure 6-7
Pivoting tray icon associations

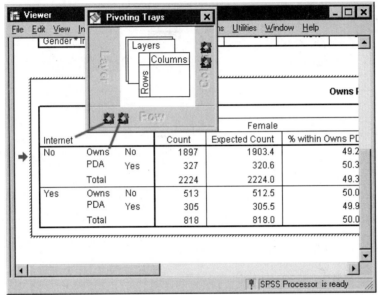

▶ Drag the *Owns PDA* icon to the left of the *Internet* icon and release the mouse button to reverse the order of these two rows

Creating and Displaying Layers

Layers can be useful for large tables with nested categories of information. By creating layers, you simplify the look of the table, making it easier to read. Layers work best when the table has at least three variables.

▶ Double-click the *Owns PDA * Gender * Internet Crosstabulation* table.

▶ If the Pivoting Trays window is not visible, from the menus choose:
Pivot
  Pivoting Trays

Working with Output

▶ Drag the *Gender* pivot icon to the Layer dimension.

Figure 6-8
Gender pivot icon in the Layer dimension

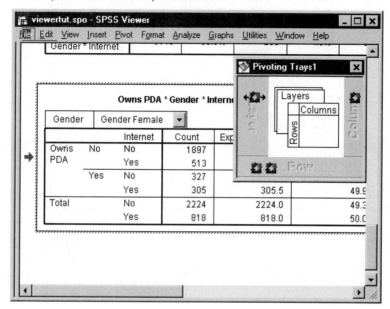

To view the different layers, you can either click the arrows on the layer pivot icon, or you can select a layer from the drop-down list in the table.

Chapter 6

Figure 6-9
Choosing a layer

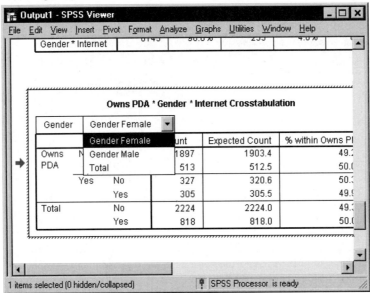

Editing Tables

Unless you've taken the time to create a custom TableLook, pivot tables are created with standard formatting. You can change the formatting of any text within a table. Formats that you can change include font name, font size, font style (bold or italic), and color.

▶ Double-click the *Level of education* table.

▶ If the Formatting toolbar is not visible, from the menus choose:
View
  Toolbar

▶ Click the title text, *Level of education*.

▶ From the drop-down list of font sizes on the toolbar, select 12.

▶ To change the color of the title text, click the Text Color tool and select a color.

Figure 6-10
Reformatted title text in the pivot table

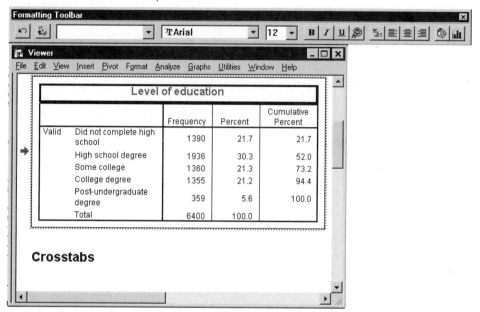

You can also edit the contents of tables and labels. For example, you can change the title of this table.

▶ Double-click the title.

▶ Type Education Level for the new label.

Note that if you change the values in a table, totals and other statistics are not recalculated.

Chapter 6

Hiding Rows and Columns

Some of the data displayed in a table may not be useful or it may unnecessarily complicate the table. Fortunately, you can hide entire rows or columns without losing any data.

▶ Double-click the *Education Level* table.

▶ Ctrl-Alt-click on the *Valid Percent* column label to select all of the cells in that column.

▶ Right-click the highlighted column and select Hide Category from the pop-up context menu.

The column is now hidden but not deleted.

To redisplay the column:

▶ From the menus choose:
View
  Show All

Rows can be hidden and displayed in the same way as columns.

Changing Data Display Formats

You can easily change the display format of data in pivot tables.

▶ Double-click the *Education Level* table.

▶ Click on the *Percent* column label to select it.

▶ From the menus choose:
Edit
  Select
    Data Cells

Working with Output

▶ From the menus choose:
Format
  Cell Properties

▶ Type 0 in the Decimals field to hide all decimal points in this column.

Figure 6-11
Cell Properties dialog box

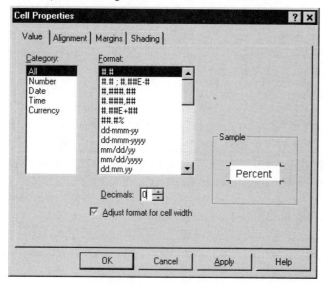

You can also change the data type and format in this dialog box.

▶ Select the type that you want from the Category list, and then select the format for that type in the Format list.

Chapter 6

▶ Click OK to apply your changes and return to the Viewer.

Figure 6-12
Decimals hidden in Percent column

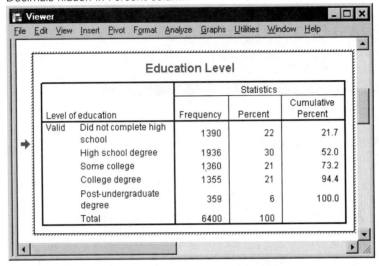

The decimals are now hidden in the *Percent* column.

TableLooks

The look and feel of your tables are a critical part of providing clear, concise, and meaningful results. For example, if your table is difficult to read, the information contained within that table may not be easily understood.

Using Predefined Formats

▶ Double-click the *Marital status* table.

▶ From the menus choose:
Format
  TableLooks...

Working with Output

The TableLooks dialog box lists a variety of predefined styles. Select a style from the list to preview it in the Sample window to the right.

Figure 6-13
TableLooks dialog box

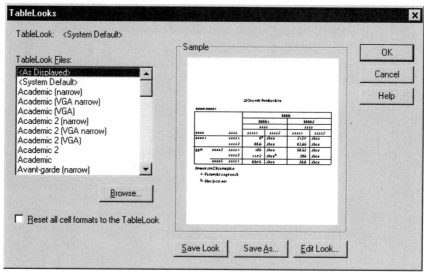

You can use a style as is, or you can edit an existing style to better suit your needs.

▶ To use an existing style, select one and click OK.

Customizing TableLook Styles

You can customize a format to fit your specific needs. Almost all aspects of a table can be customized, from the background color to the border styles.

▶ Double-click the *Marital status* table.

▶ From the menus choose:
Format
　TableLooks...

Chapter 6

▶ Select the style that is closest to your desired format and click Edit Look.

▶ Click the Cell Formats tab to view the formatting options.

Figure 6-14
Table Properties dialog box

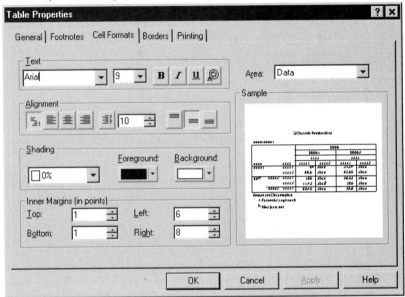

The formatting options include font name, font size, style, and color. Additional options include alignment, shading, foreground and background colors, and margin sizes.

The Sample window provides a preview of how the formatting changes affect your table. Each area of the table can have different formatting styles. For example, you probably wouldn't want the title to have the same style as the data. To select a table area to edit, you can either choose the area by name in the Area drop-down list, or you can click the area that you want to change in the Sample window.

▶ Select Title from the Area drop-down list.

▶ Select a new color from the Background drop-down list.

The Sample window shows the new style.

Working with Output

▶ Click OK to return to the TableLooks dialog box.

You can save your new style, which allows you to apply it to future tables easily.

▶ Click Save As.

▶ Navigate to the desired target directory and enter a name for your new style in the File Name text box.

▶ Click Save.

▶ Click OK to apply your changes and return to the Viewer.

The table now contains the custom formatting that you specified.

Figure 6-15
Custom TableLook

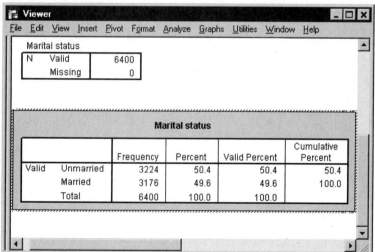

Changing the Default Table Formats

Although you can change the format of a table after it has been created, it may be more efficient to change the default TableLook so that you do not have to change the format every time you create a table.

Chapter 6

To change the default TableLook style for your pivot tables, from the menus choose:

Edit
  Options...

▶ Click the Pivot Tables tab in the Options dialog box.

Figure 6-16
Options dialog box

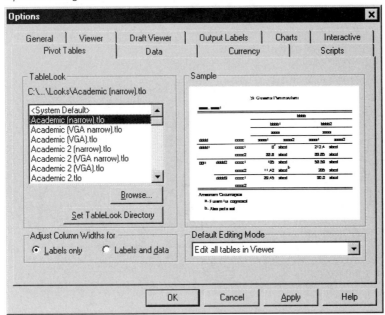

▶ Select the TableLook style that you want to use for all new tables.

Use the Sample window to see a preview of each TableLook.

▶ Click OK to save your settings and close the dialog box.

All tables that you create after changing the default TableLook automatically conform to the new formatting rules.

Working with Output

Customizing the Initial Display Settings

The initial display settings include the alignment of objects in the Viewer, whether objects are shown or hidden by default, and the width of the Viewer window. To change these settings:

▶ From the menus choose:
Edit
  Options...

▶ Click the Viewer tab.

Figure 6-17
Viewer options

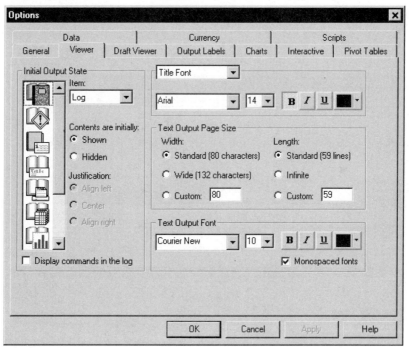

The settings are applied on an object-by-object basis. For example, you can customize the way charts are displayed without making any changes to the way tables are displayed. Simply select the object that you want to customize, and make the desired changes.

▶ Click the Title icon to display its settings.

▶ Click Center to display all titles in the (horizontal) center of the Viewer.

You can also hide elements, such as the log and warning messages, that tend to clutter your output. Double-clicking on an icon automatically changes that object's display property.

▶ Double-click the Warnings icon to hide warning messages in the output.

▶ Click OK to save your changes and close the dialog box.

Figure 6-18
Centered title in the Marital status table

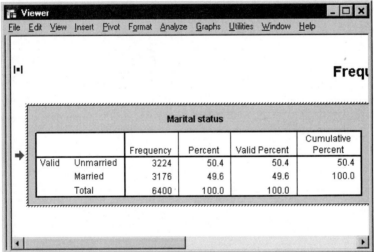

Your new settings will be applied the next time you run a statistical procedure. All items that you have hidden will still be created, but they will not be visible in the contents pane. Centered items can be identified by the small symbol to the left of the item.

Working with Output

Displaying Variable and Value Labels

In most cases, displaying the labels for variables and values is more effective than displaying the variable name or the actual data value. There may be cases, however, when you want to display both the names and the labels.

▶ From the menus choose:
Edit
   Options...

▶ Click the Output Labels tab.

Figure 6-19
Output Labels options

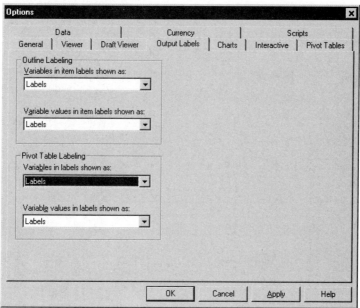

You can specify different settings for the outline and contents panes. For example, to show labels in the outline and variable names and data values in the contents:

Chapter 6

▶ In the Pivot Table Labeling group, select **Names** from the Variables in Labels drop-down list to show variable names instead of labels.

▶ Then, select **Values** from the Variable Values in Labels drop-down list to show data values instead of labels.

Figure 6-20
Pivot Table Labeling settings

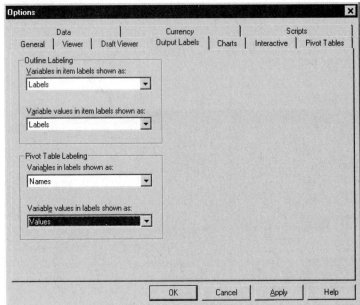

Working with Output

The new settings are applied the next time you run a statistical procedure.

Figure 6-21
Variable names and values displayed

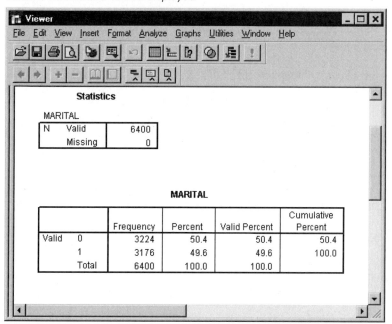

Using Results in Other Applications

Your results can be used in many applications. For example, you may want to include a chart or graph in a presentation or report. Applications such as Microsoft's PowerPoint or Word can display your results as plain text, rich text, or as a metafile, which is a graphical representation of the output.

The following examples are specific to Microsoft Word, but they may work similarly in other word processing applications.

Pasting Results as Rich Text

You can paste pivot tables into Word as native Word tables. Text formatting, such as font size and color, is not retained, but columns and rows are properly aligned. Because the table is in a text format, the data can be edited after you paste it into your document.

▶ Click the *Marital status* table in the Viewer.

▶ From the menus choose:
Edit
  Copy

▶ Open your word processing application.

▶ From the word processor's menus choose:
Edit
  Paste Special

▶ Select Formatted Text (RTF) in the Paste Special dialog box.

Figure 6-22
Paste Special dialog box

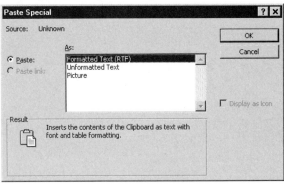

Working with Output

▶ Click OK to paste your results into the current document.

Figure 6-23
Pivot table displayed in Word

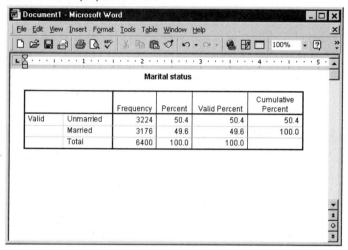

The table is now displayed in your document. You can apply custom formatting, edit the data, and resize the table to fit your needs.

Pasting Results as Metafiles

Pasting your results as metafiles maintains the formatting of your output. Your pasted output becomes an image in the target document.

▶ Click the *Marital status* table in the Viewer.

▶ From the menus choose:
Edit
  Copy

Chapter 6

- ▶ Open your word processing application.

- ▶ From the word processor's menus choose:
  Edit
    Paste Special

- ▶ Select Picture in the Paste Special dialog box. (In some applications, the choice may be "metafile" instead of "Picture.")

  Figure 6-24
  Paste Special dialog box

  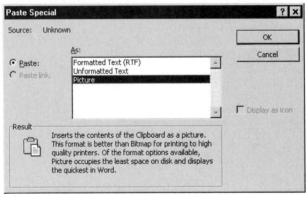

- ▶ Click OK to paste your results into the current document.

Working with Output

Figure 6-25
Metafile displayed in Word

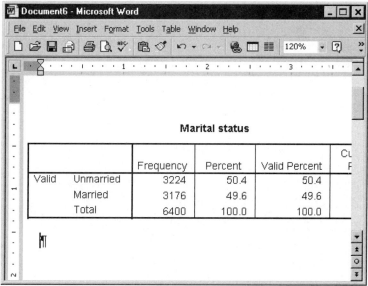

The metafile is now embedded in your document. This image is a snapshot of the *Marital status* table. Only the visible portions of the table are copied. Information in hidden categories or layers is not included in the metafile.

Pasting Results as Text

Pivot tables can be copied to other applications as plain text. Formatting styles are not retained in this method, but you gain the ability to edit the table data after you paste it into the target application.

▶ Click the *Marital status* table in the Viewer.

▶ From the menus choose:
Edit
  Copy

▶ Open your word processing application.

▶ From the word processor's menus choose:
Edit
  Paste Special

▶ Select Unformatted Text in the Paste Special dialog box.

Figure 6-26
Paste Special dialog box

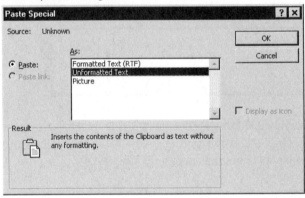

▶ Click OK to paste your results into the current document.

Working with Output

Figure 6-27
Pivot table displayed in Word

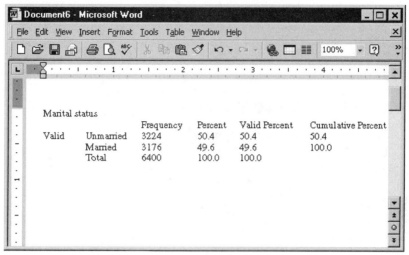

Each column of the table is separated by tabs. You can change the column widths by adjusting the tab stops in your word processing application.

Exporting Results to Microsoft Word and Excel Files

SPSS allows you to export results to a single Microsoft Word or Excel file. You can export selected items or all items in the Viewer, and, if exporting to Word, you can export charts as embedded Windows metafiles. This topic uses the files *msouttut.spo* and *demo.sav*.

In the Viewer's outline pane, you can select specific items that you want to export. You do not have to select specific items.

Chapter 6

▶ From the Viewer menus choose:
File
 Export...

There are several options for exporting the results.

Figure 6-28
Export Output dialog box

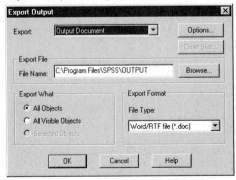

First, you can select which type of output file you want to create. For Word, you can create a file that contains charts (Output Document) or one that does not contain charts (Output Document (No Charts)). Charts are embedded in the Word document as Windows metafiles. For Excel, you can create only documents that do not contain charts (Output Document (No Charts)).

▶ Select Output Document from the Export drop-down list.

You can save the exported Word or Excel file to any location and assign it any name that Windows allows. The default file name is *OUTPUT*, and the default path is the installation location or, if you opened or saved an output file, the location of that file. You can change the default by typing a new path or clicking Browse to locate

Working with Output

a destination. You do not have to specify a file extension. The export functionality adds the appropriate extension.

Note the default file path in the File Name text box. You will need to know where the file is to open it in Word.

Instead of exporting all objects in the Viewer, you can choose to export only visible objects (open books in the outline pane) or those that you selected in the outline pane. If you did not select any items in the outline pane, you do not have the option to export selected objects.

▶ Select All Objects in the Export What group.

Finally, you select the file format.

▶ Select Word/RTF file (*.doc) from the File Type drop-down list.

▶ Click OK to generate the Word file.

When you open the resulting file in Word, you can see how the results are exported. Notes, which are not visible objects, appear in Word because you chose to export all objects.

Figure 6-29
Output.doc in Word

Pivot tables become Word tables, with all of the formatting of the original pivot table, including fonts, colors, borders, and so on.

Figure 6-30
Pivot tables in Word

		Frequency	Percent	Valid Percent	Cumul Perce
Valid	Female	3179	49.7	49.7	
	Male	3221	50.3	50.3	
	Total	6400	100.0	100.0	

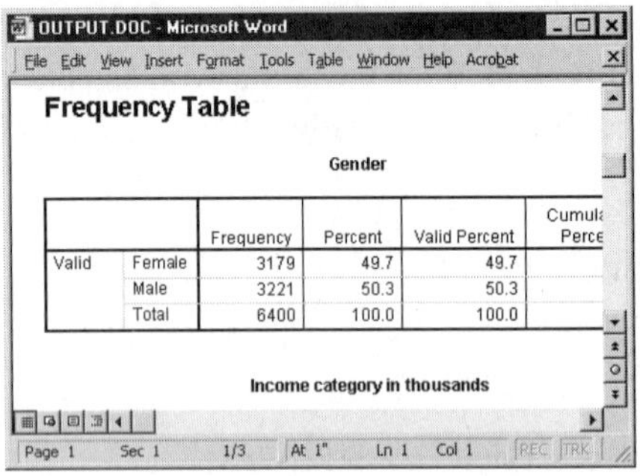

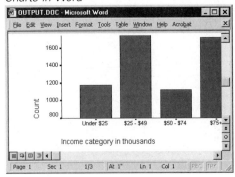

Charts become embedded Windows metafiles.

Figure 6-31
Charts in Word

Text output is displayed in a fixed-pitch font.

Figure 6-32
Text output in Word

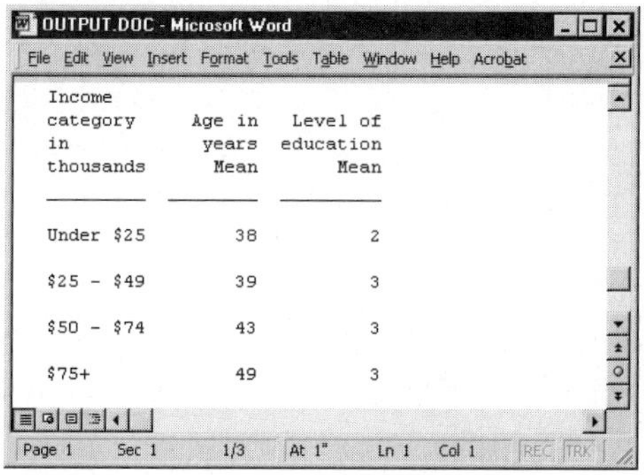

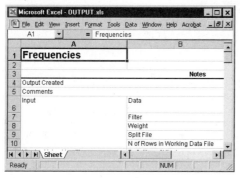

If you export to an Excel file, results are exported differently.

Figure 6-33
Output.xls in Excel

Chapter 6

Pivot table rows, columns, and cells become Excel rows, columns, and cells.

Figure 6-34
Pivot tables in Excel

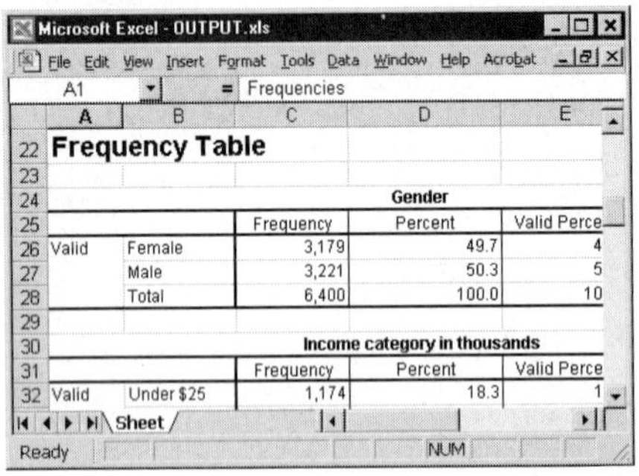

Each line in the text output is a row in the Excel file, with the entire contents of the line contained in a single cell. Charts are not exported at all.

Figure 6-35
Text output in Excel

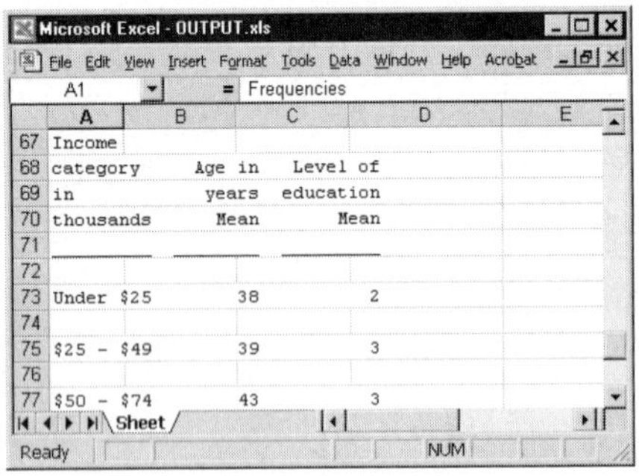

Working with Output

Exporting Results to HTML and Text Formats

You can also export results to HTML (hypertext markup language) and text formats and a number of graphic formats. When saving as HTML and text, all non-graphic output from SPSS can be exported into a single file and read by other programs. HTML format is commonly used for intranet and Internet Web pages, while text can be read by all word processors. In this way, SPSS output can be easily incorporated into documents on computers that do not run SPSS or onto Web pages.

When you export HTML and text output, charts can be exported as well, but not to a single file.

Figure 6-36
Chart in HTML

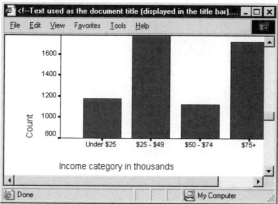

Each chart will be saved as a file in a format that you specify, and references to these graphics files will be placed in the HTML or text document created by SPSS. There is also an option to export all charts, or selected charts, in separate graphics files.

Figure 6-37
References to graphics in HTML output

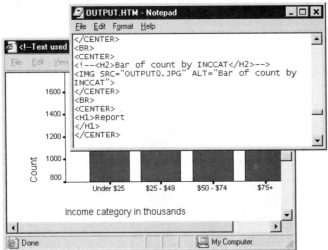

Chapter 7

Creating and Editing Charts

A wide variety of chart types are available, and many of those charts are available in two different formats:

- **Standard charts.** Charts created from the main Graphs menu and charts created by statistical procedures.
- **Interactive charts.** Charts created from the Interactive submenu of the Graphs menu and charts created from pivot tables.

The examples in this chapter use the data file *demo.sav*.

Creating Standard Charts

In this example, we'll create a simple pie chart that shows how many respondents have Internet service at home.

▶ From the menus choose:
Graphs
  Pie...

Chapter 7

▶ Click Summaries for groups of cases and then click Define.

Figure 7-1
Pie Charts dialog box

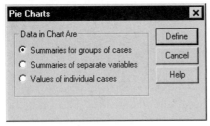

Since we want to base the chart on a single variable, we selected Summaries for groups of cases. Chart elements (bars, pie slices) can also be based on summaries of separate variables or values from individual cases in the data file.

▶ Select *Internet* as the variable that defines slices (Define Slices by).

Figure 7-2
Define Pie dialog box

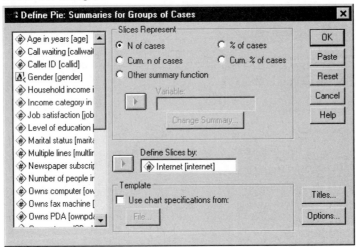

When charts are created, they do not show the *missing* category by default. You want to display this category to make sure that the number of cases with missing values is not excessive.

Creating and Editing Charts

▶ Click Options.

▶ Select Display groups defined by missing values, and then click Continue.

Figure 7-3
Options dialog box

▶ Click OK in the Define Pie dialog box to create the pie chart.

Figure 7-4
Pie chart

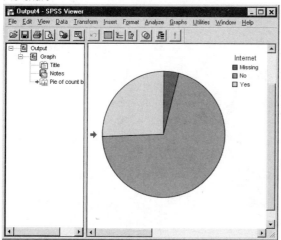

The pie chart reveals that most respondents do not have Internet service at home. From the chart, it appears that only about a quarter of the respondents have Internet service.

Chapter 7

Editing Standard Charts

You can edit charts in a variety of ways. For the sample pie chart we created, we will:

- Add a title.
- Remove the small category of missing data.
- Display percentages for the two remaining categories in the chart.

The first thing we'll do is add a title.

▶ Double-click the pie chart to open it in the Chart Editor.

Figure 7-5
Pie chart in the Chart Editor

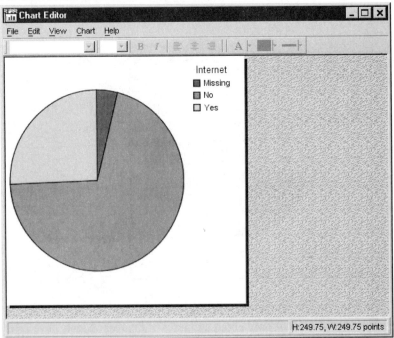

The pie chart fills up all of the available space, leaving no room for the title. Therefore, before adding the title, we need to resize the data frame. The data frame is the frame around the pie chart.

▶ Select the data frame. When it is selected, you will see selection handles for the frame.

▶ Position the mouse cursor over the bottom right resizing handle so that a two-headed arrow appears.

▶ Click while pressing Shift, and drag the frame to resize it. Pressing and holding Shift maintains the frame's aspect ratio, which is the ratio of width to height.

After resizing the frame, you need to move it to make room for the title text box.

▶ Position the mouse cursor over the frame's border. When a four-headed arrow appears, click and drag the frame to a new location that provides space for the title text box.

Now you can add the title.

▶ From the Chart Editor menus choose:
Chart
  Add Chart Element
    Text Box

▶ Type Home Internet Service into the text box, and then press Enter.

Chapter 7

A descriptive title is now displayed above the pie chart. You can move the text box if desired.

Figure 7-6
Pie chart with title

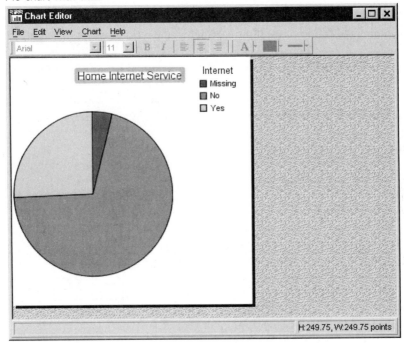

Now we'll remove the small category of missing data.

▶ Select the pie chart.

▶ From the Chart Editor menus choose:
Edit
  Properties

▶ In the Properties window, click the Categories tab.

▶ Move *Missing* from the Order list to the Excluded list.

Figure 7-7
Categories tab

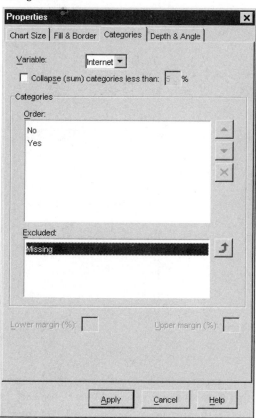

▶ Click Apply.

The category of missing data has been removed from the pie chart, leaving only two categories.

Figure 7-8
Pie chart without missing category

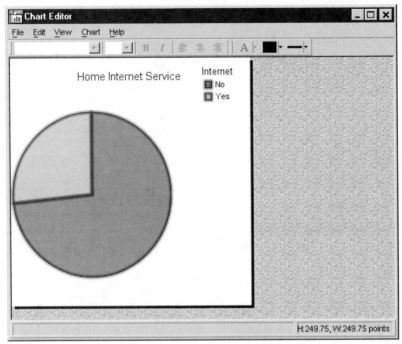

The pie chart clearly shows that most respondents do not have Internet service at home, and it looks like almost three-quarters of the respondents are in the *No* category. However, it might be useful to see the exact percentages.

▶ Select the pie chart.

▶ From the menus choose:
Chart
  Show Data Labels

Now the pie chart displays labels of counts. We need to change the labels to percentages.

Figure 7-9
Pie chart with counts

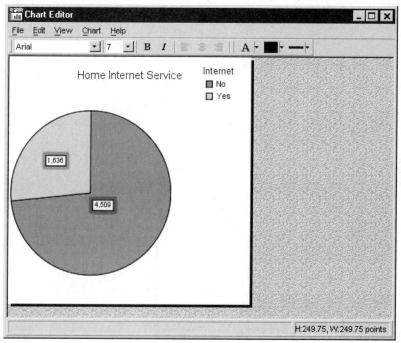

▶ On the Data Value Labels tab, move *Count* from the Contents list to the Available list.

Chapter 7

▶ Move *Percent* from the Available list to the Contents list.

Figure 7-10
Data Value Labels tab

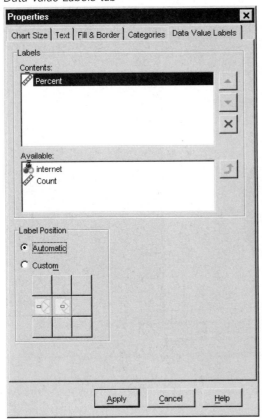

▶ Click Apply.

Creating and Editing Charts

Now percentages are displayed in the pie slices.

Figure 7-11
Pie chart with percentages

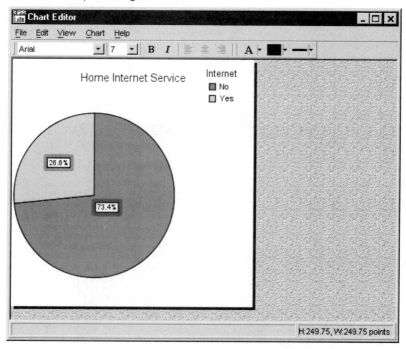

The percentages are based on the two categories displayed (73.4 + 26.6 = 100). If you put the category containing missing values back into the pie, the percentages will change.

Creating Interactive Charts

Interactive charts are created by selecting a chart type from the Interactive submenu of the Graphs menu.

Chapter 7

▶ From the menus choose:
   Graphs
     Interactive
       Bar...

▶ Drag and drop *Owns PDA (ownpda)* into the *x* (horizontal) axis list.

▶ Drag and drop *Income category in thousands (inccat)* into the Color list.

Figure 7-12
Create Bar Chart dialog box

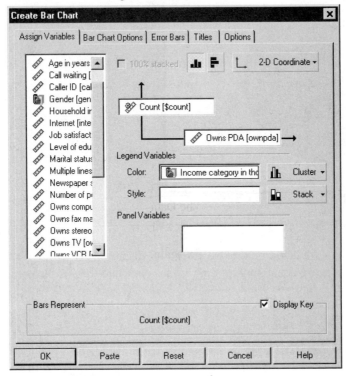

▶ Click OK to create the chart.

Creating and Editing Charts

Figure 7-13
Interactive bar chart

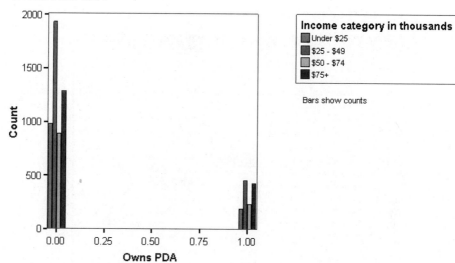

You might notice that this is not a particularly attractive chart. That's because interactive graphs treat scale and categorical variables differently, and *Owns PDA (ownpda)* is defined as a scale variable. Since this is really a categorical variable, we could change the variable definition in the Data Editor. Or we could simply tell the interactive graphics procedure to treat it as a categorical variable.

▶ Open the interactive bar chart dialog box again.

▶ Right-click on *Owns PDA (ownpda)* on the *x* (horizontal) axis list.

Chapter 7

▶ Select Categorical from the pop-up context menu, and then click OK to create the chart.

Figure 7-14
Variable type context menu

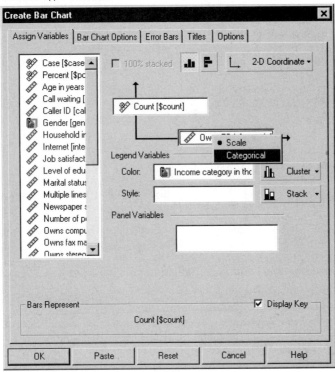

Now the clustered bar chart looks much better.

Figure 7-15
Interactive bar chart

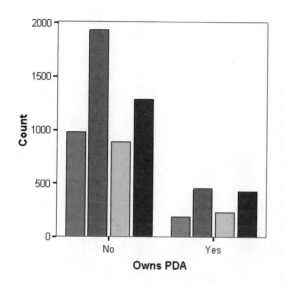
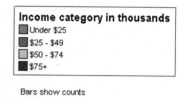

Editing Interactive Charts

You can modify both interactive and standard charts; however, interactive charts can be edited directly in the Viewer.

▶ Double-click the chart to activate it.

Interactive charts are activated and edited in place in the Viewer window, while standard charts open in a separate window for editing.

Figure 7-16
Activated interactive chart

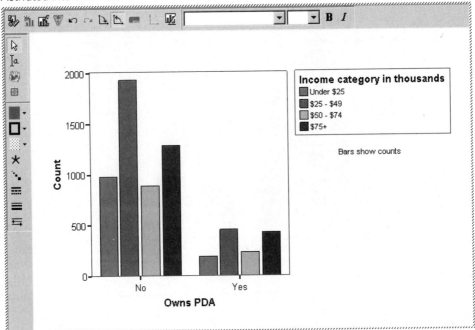

To change the color of a bar, you simply click a bar to select it, and then select another color from the Fill Color palette on the vertical toolbar. However, in this clustered bar chart example, colors are associated with pairs of bars, so you change the color by selecting the bar category in the legend.

▶ Click the color square next to the $25–$49 category in the legend.

Creating and Editing Charts

▶ Click the down arrow next to the Fill Color icon on the vertical toolbar, and select a new color.

Figure 7-17
Fill Color palette

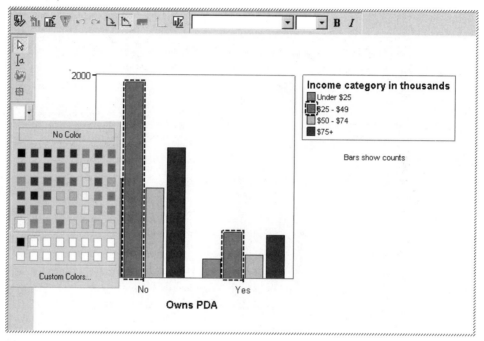

The color is applied to the two bars that represent the $25–$49 category.

Chapter 7

In this example, the key text below the legend is unnecessary since the scale axis is already labeled *Counts*. To delete the key text:

▶ Right-click on the text and select Hide key from the pop-up context menu.

Figure 7-18
Right-mouse context menu

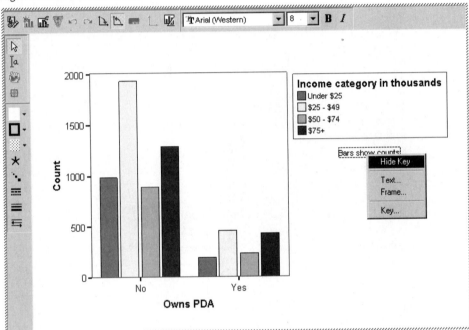

Creating and Editing Charts

Figure 7-19
Key text hidden below legend

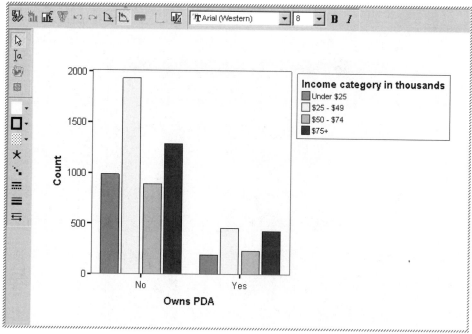

The key text is now hidden.

Creating an Interactive Chart from a Pivot Table

You can create interactive charts from data contained in a pivot table but first you need a pivot table.

▶ From the menus choose:
Analyze
  Descriptive Statistics
    Crosstabs...

Chapter 7

▶ Select *Income category in thousands (inccat)* as the row variable.

▶ Select *Owns PDA (ownpda)* as the column variable.

Figure 7-20
Crosstabs dialog box

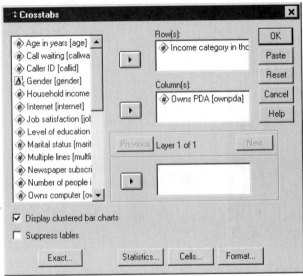

▶ Click OK to run the procedure.

▶ Double-click the pivot table to activate it.

Creating and Editing Charts

▶ Click and drag to select the data cells you want to use in the chart.

Figure 7-21
Data selected in pivot table

Income category in thousands * Owns PDA Crosstabulation			
Statistics	Count		

		Owns PDA		
		No	Yes	Total
Income category in thousands	Under $25	983	191	1174
	$25 - $49	1933	455	2388
	$50 - $74	889	231	1120
	$75+	1288	430	1718
Total		5093	1307	6400

▶ Right-click anywhere in the selected area.

▶ From the pop-up context menu choose:
Create Graph
  Bar

Chapter 7

An interactive chart of the selected data is created.

Figure 7-22
Clustered bar chart created from pivot table

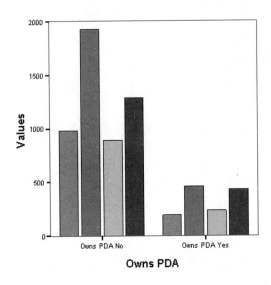
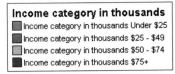

Chapter 8

Working with Syntax

SPSS syntax provides a method for you to control the product without navigating through dialog boxes, viewers, or data editors. Instead, you control the application through syntax-based commands. Nearly every action you can achieve through the user interface can be achieved through syntax. Using syntax also allows you to save the exact specifications used during a session. Syntax is not available with the Student Version.

The examples in this chapter use the data file *demo.sav*.

Pasting Syntax

The easiest way to create syntax is to use the Paste button located on most dialog boxes.

▶ Open *demo.sav* for use in this example.

▶ From the menus choose:
Analyze
  Descriptive Statistics
    Frequencies...

This opens the Frequencies dialog box.

Figure 8-1
Frequencies dialog box

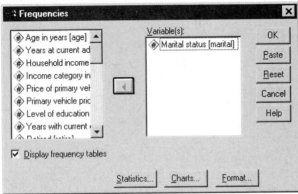

- Select *Marital status (marital)* in the source list.

- Click the arrow button to move the variable to the Variables list.

- Click Charts.

- In the Charts dialog box, select Bar charts.

- In the Chart Values group, select Percentages.

- Click Continue.

- Click Paste to copy the syntax created as a result of the dialog box selections to the Syntax Editor.

Working with Syntax

Figure 8-2
Frequencies syntax

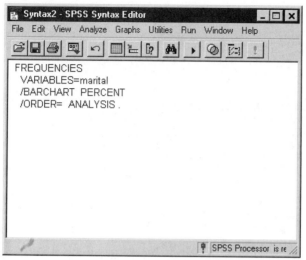

You can use this syntax alone, add it to a larger syntax file, or refer to it in a Production Facility job.

▶ To run the syntax currently displayed, from the menus choose:
Run
  Current

Chapter 8

Editing Syntax

In the syntax window, you can edit the syntax. For example, you could change the subcommand /BARCHART to display frequencies instead of percentages. (A subcommand is indicated by a slash.)

Figure 8-3
Modified syntax

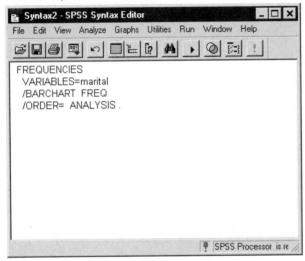

To find out what subcommands and keywords are available for the current command, click the Syntax Help button. The example shows the complete syntax for the FREQUENCIES command.

Working with Syntax

Figure 8-4
FREQUENCIES syntax help

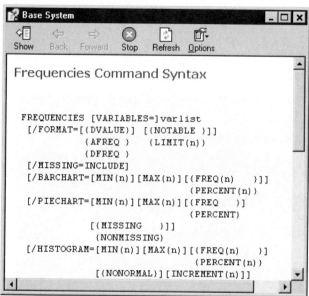

If the cursor is not in a command, clicking the Syntax Help button displays an alphabetical list of commands. You can select the one you want.

Typing Syntax

You can type syntax into a syntax window that is already open, or you can open a new syntax window by choosing:
File
 New
  Syntax...

Chapter 8

Saving Syntax

To save a syntax file, from the menus choose:
File
  Save

or

File
  Save As...

Either action opens the standard dialog box for saving files.

Opening and Running a Syntax File

▶ To open a saved syntax file, from the menus choose:
File
  Open
    Syntax...

▶ Select a syntax file. If no syntax files are displayed, make sure Syntax (*.sps) is selected in the Files of type drop-down list.

▶ Click Open.

▶ To run the syntax currently displayed, from the menus choose:
Run
  Current

If the commands apply to a specific data file, the data file must be opened before running the commands, or you must include a command that opens the data file. You can paste this type of command from the dialog boxes that open data files.

Chapter 9

Modifying Data Values

The data you start with may not always be organized in the most useful manner for your analysis or reporting needs. You may, for example, want to:

- Create a categorical variable from a scale variable.
- Combine several response categories into a single category.
- Create a new variable that is the computed difference between two existing variables.

This chapter uses the data file *demo.sav*.

Creating a Categorical Variable from a Scale Variable

Several categorical variables in the data file *demo.sav* are, in fact, derived from scale variables in that data file. For example, the variable *inccat* is simply *income* grouped into four categories. This categorical variable uses the integer values 1–4 to represent the following income categories: less than 25, 25–49, 50–74, and 75 or higher.

To create the categorical variable *inccat*:

▶ From the menus in the Data Editor window choose:
Transform
  Visual Bander...

Chapter 9

Figure 9-1
Initial Visual Bander dialog box

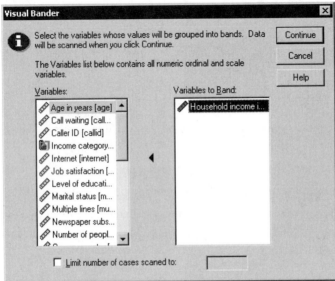

In the initial Visual Bander dialog box, you select the scale and/or ordinal variables for which you want to create new, banded variables. **Banding** means taking two or more contiguous values and grouping them into the same category.

Since the Visual Bander relies on actual values in the data file to help you make good banding choices, it needs to read the data file first. Since this can take some time if your data file contains a large number of cases, this initial dialog box also allows you to limit the number of cases to read ("scan"). This is not necessary for our sample data file. Even though it contains more than 6,000 cases, it does not take long to scan that number of cases.

▶ Drag and drop *Household income in thousands [income]* from the Variables list to the Variables to Band list, and then click Continue.

Modifying Data Values

Figure 9-2
Main Visual Bander dialog box

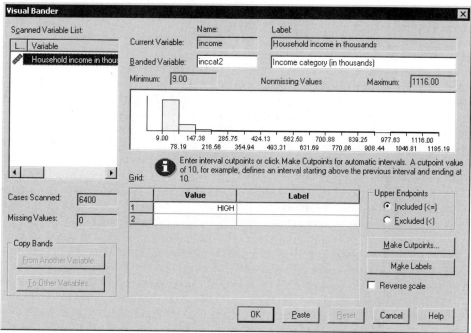

▶ In the main Visual Bander dialog box, select *Household income in thousands [income]* in the Scanned Variable List.

A histogram displays the distribution of the selected variable (which in this case is highly skewed).

▶ Enter inccat2 for the new banded variable name and Income category (in thousands) for the variable label.

▶ Click Make Cutpoints.

Figure 9-3
Visual Bander Cutpoints dialog box

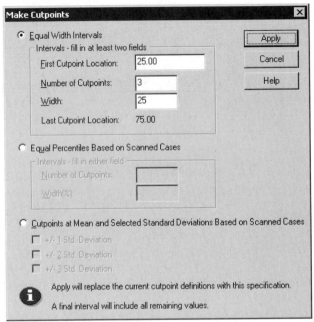

▶ Click Equal Width Intervals.

▶ Enter 25 for the first cutpoint location, 3 for the number of cutpoints, and 25 for the width.

The number of banded categories is the number of cutpoints *plus one*. So, in this example, the new banded variable will have four categories, with the first three categories each containing ranges of 25 (thousand) and the last one containing all values above the highest cutpoint value (75).

▶ Click Apply.

Modifying Data Values

Figure 9-4
Main Visual Bander dialog box with defined cutpoints

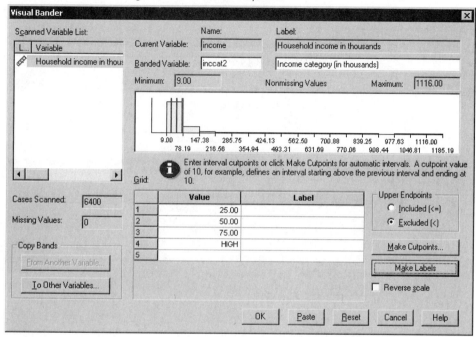

The values now displayed in the grid represent the defined cutpoints, which are the upper endpoints of each category. Vertical lines in the histogram also indicate the locations of the cutpoints.

By default, these cutpoint values are included in the corresponding categories. For example, the first value of 25 would include all values less than or equal to 25. But in this example, we want categories that correspond to less than 25, 25–49, 50–74, and 75 or higher.

▶ In the Upper Endpoints group, select Excluded (<).

▶ Then click Make Labels.

Figure 9-5
Automatically generated value labels

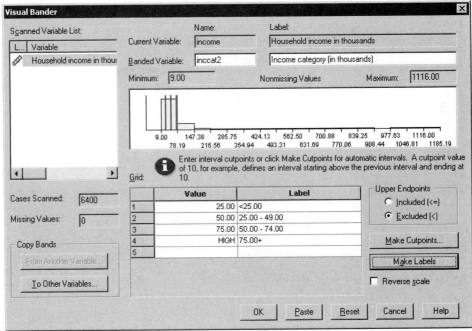

This automatically generates descriptive value labels for each category. Since the actual values assigned to the new banded variable are simply sequential integers starting with 1, the value labels can be very useful.

You can also manually enter or change cutpoints and labels in the grid, change cutpoint locations by dragging and dropping the cutpoint lines in the histogram, and delete cutpoints by dragging cutpoint lines off the histogram.

▶ Click OK to create the new, banded variable.

Modifying Data Values

The new variable is displayed in the Data Editor. Since the variable is added to the end of the file, it is displayed in the far right column in Data View and in the last row in Variable View.

Figure 9-6
New variable displayed in Data Editor

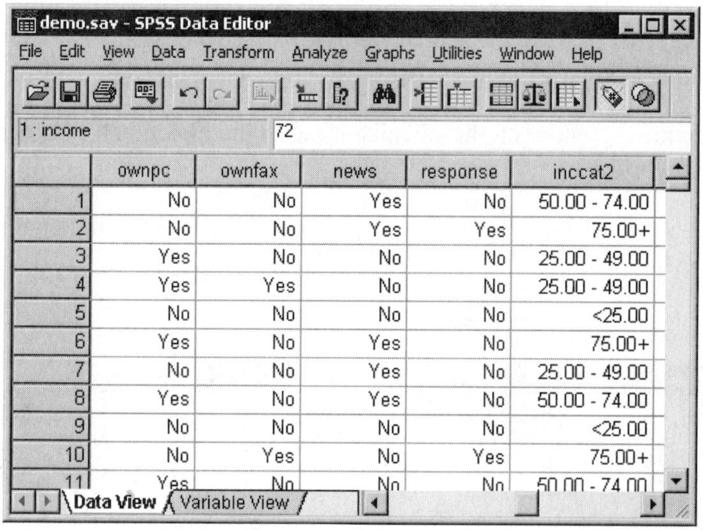

Computing New Variables

You can compute new variables based on highly complex equations, using a wide variety of mathematical functions. In this example, however, we will simply compute a new variable that is the difference between the values of two existing variables.

The data file *demo.sav* contains a variable for the respondent's current age and a variable for the number of years at current job. It does not, however, contain a variable for the respondent's age at the time he or she started that job. We can create a new variable that is the computed difference between current age and number of years at current job, which should be the approximate age at which the respondent started that job.

Chapter 9

▶ From the menus in the Data Editor window choose:
Transform
    Compute...

▶ For Target Variable, enter jobstart.

▶ Select *Age in years (age)* in the source variable list and click the arrow button to copy it to the Numeric Expression text box.

▶ Click the minus (–) button on the calculator pad in the dialog box (or the minus key on the keyboard).

▶ Select *Years with current employer (employ)* and click the arrow button to copy it to the expression.

Figure 9-7
Compute Variable dialog box

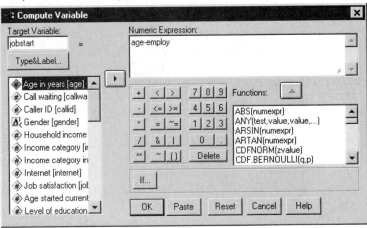

Note: Be careful to select the correct employment variable. There is also a recoded categorical version of the variable, which is *not* what you want. The numeric expression should be *age-employ*, not *age-empcat*.

▶ Click OK to compute the new variable.

Modifying Data Values

The new variable is displayed in the Data Editor. Since the variable is added to the end of the file, it is displayed in the far right column in Data View and in the last row in Variable View.

Figure 9-8
New variable displayed in the Data Editor

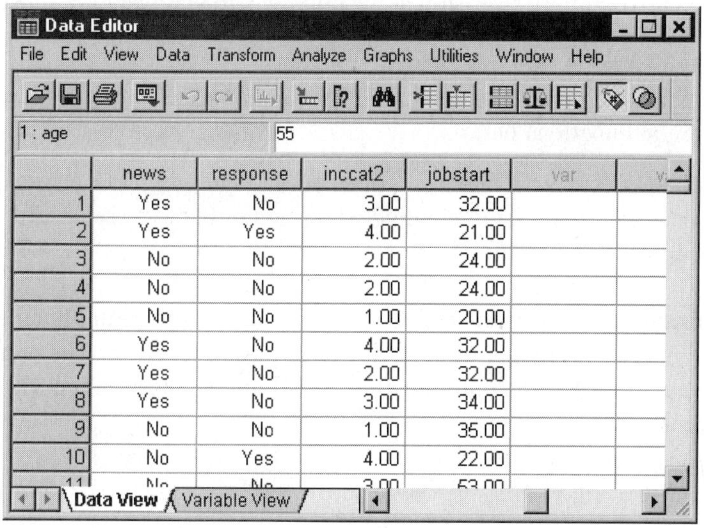

Using Functions in Expressions

You can also use predefined functions in expressions. The function list contains more than 70 built-in functions, including:

- Arithmetic functions
- Statistical functions
- Distribution functions
- Logical functions
- Date and time aggregation and extraction functions
- Missing-value functions
- Cross-case functions
- String functions

For information on specific functions, right-click the function you want to know about or click Help in the Compute Variable dialog box.

Pasting a function into an expression. To paste a function into an expression:

▶ Position the cursor in the expression at the point where you want the function to appear.

▶ Double-click the function in the Functions list (or select the function and click the arrow above the Functions list).

The function is inserted into the expression. If you highlight part of the expression and then insert the function, the highlighted portion of the expression is used as the first argument in the function.

Editing a function in an expression. The function is not complete until you enter the arguments, represented by question marks in the pasted function. The number of question marks indicates the minimum number of arguments required to complete the function.

▶ Highlight the question mark(s) in the pasted function.

▶ Enter the arguments. If the arguments are variable names, you can paste them from the variable list.

Using Conditional Expressions

You can use conditional expressions (also called logical expressions) to apply transformations to selected subsets of cases. A conditional expression returns a value of true, false, or missing for each case. If the result of a conditional expression is true, the transformation is applied to that case. If the result is false or missing, the transformation is not applied to the case.

To specify a conditional expression:

▶ Click If in the Compute Variable dialog box.

Modifying Data Values

This opens the If Cases dialog box.

Figure 9-9
If Cases dialog box

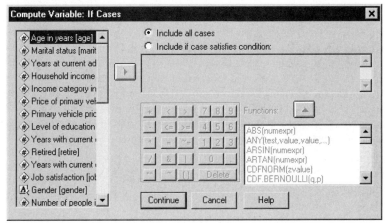

▶ Select Include if case satisfies condition.

▶ Enter the conditional expression.

Most conditional expressions contain at least one relational operator, as in:

age>=21

or

income*3<100

In the first example, only cases with a value of 21 or greater for *Age (age)* are selected. In the second example, *Household income in thousands (income)* multiplied by 3 must be less than 100 for a case to be selected.

You can also link two or more conditional expressions using logical operators, as in:

age>=21 | ed>=4

or

income*3<100 & ed=5

Chapter 9

In the first example, cases that meet either the *Age (age)* condition or the *Level of education (ed)* condition are selected. In the second example, both the *Household income in thousands (income)* and *Level of education (ed)* conditions must be met for a case to be selected.

Computing Variables with Missing Values

When computing variables with missing values, it's a good idea to perform a trial on a small data file in order to ensure that your missing data is handled in a way that is acceptable. Some computations handle missing data by removing them from the equation, while others cannot perform the computation and the new value is considered missing. The following examples use *car_sales.sav*.

There are two different ways to compute the mean of a set of variables. The first is to divide the sum of the set of variables by the number of variables in the set.

▶ From the menus choose:
Transform
  Compute...

▶ Type avgsale in the Target Variable text box.

▶ Type (resale + price) / 2 in the Numeric Expression text box.

Figure 9-10
Find the average

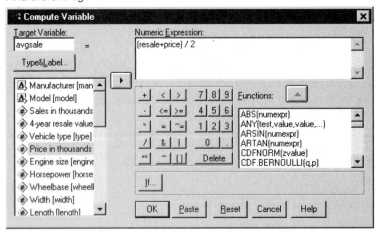

Modifying Data Values

This equation finds the average of the original price of a car and its four-year resale price.

▶ Click OK.

A new variable has been added to the end of the data file. This variable, which we named *avgsale*, is in the far right column in Data View. Any cases that have missing values in either of the two variables that we computed also have a missing value for the *avgsale* variable.

Figure 9-11
Data Editor with avgsale added

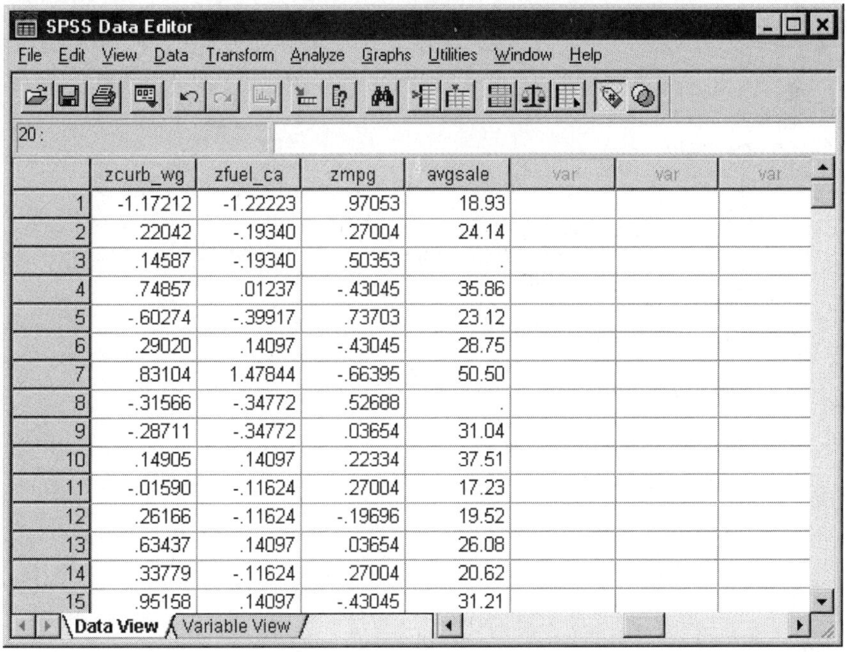

	zcurb_wg	zfuel_ca	zmpg	avgsale
1	-1.17212	-1.22223	.97053	18.93
2	.22042	-.19340	.27004	24.14
3	.14587	-.19340	.50353	.
4	.74857	.01237	-.43045	35.86
5	-.60274	-.39917	.73703	23.12
6	.29020	.14097	-.43045	28.75
7	.83104	1.47844	-.66395	50.50
8	-.31566	-.34772	.52688	.
9	-.28711	-.34772	.03654	31.04
10	.14905	.14097	.22334	37.51
11	-.01590	-.11624	.27004	17.23
12	.26166	-.11624	-.19696	19.52
13	.63437	.14097	.03654	26.08
14	.33779	-.11624	.27004	20.62
15	.95158	.14097	-.43045	31.21

The second way to compute the mean of a set of variables is to use the mean function.

▶ Recall the Compute Variable dialog box.

▶ Click Reset to clear the previous selections in this dialog box.

▶ Type **meansale** in the Target Variable text box.

Chapter 9

▶ Type mean(resale, price) in the Numeric Expression text box.

Figure 9-12
Find the mean

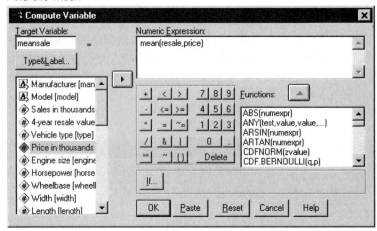

Like the previous example, this equation finds the average, or mean, of the car's original price and its four-year resale price.

▶ Click OK.

Modifying Data Values

The *meansale* variable is added to the far right column in Data View. Unlike the previous example, the mean function removes missing values from the equation and computes the mean of the remaining values. You can modify the behavior of this function to limit the number of values that can be removed from the equation.

Figure 9-13
Results from two mean calculations

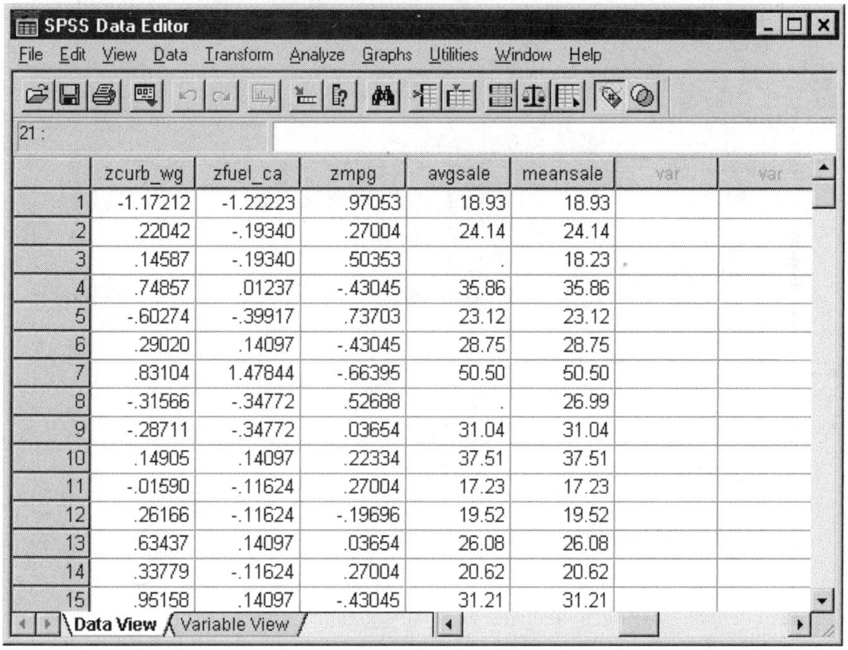

	zcurb_wg	zfuel_ca	zmpg	avgsale	meansale
1	-1.17212	-1.22223	.97053	18.93	18.93
2	.22042	-.19340	.27004	24.14	24.14
3	.14587	-.19340	.50353	.	18.23
4	.74857	.01237	-.43045	35.86	35.86
5	-.60274	-.39917	.73703	23.12	23.12
6	.29020	.14097	-.43045	28.75	28.75
7	.83104	1.47844	-.66395	50.50	50.50
8	-.31566	-.34772	.52688	.	26.99
9	-.28711	-.34772	.03654	31.04	31.04
10	.14905	.14097	.22334	37.51	37.51
11	-.01590	-.11624	.27004	17.23	17.23
12	.26166	-.11624	-.19696	19.52	19.52
13	.63437	.14097	.03654	26.08	26.08
14	.33779	-.11624	.27004	20.62	20.62
15	.95158	.14097	-.43045	31.21	31.21

If you want to find the mean of four variables and only want a result if three or more of the variables have valid values, you could use the following syntax:

newvar = mean.3(var1, var2, var3, var4)

Figure 9-14
Specifying number of valid values for mean calculation

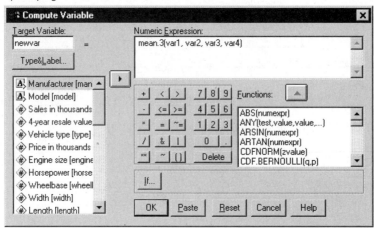

To raise or lower the number of required valid values, simply change the number after mean (which is 3 in this example).

Chapter 10

Sorting and Selecting Data

Data files are not always organized in the ideal form for your specific needs. To prepare data for analysis, you can select from a wide range of file transformations, including the ability to:

- **Sort data.** You can sort cases based on the value of one or more variables.
- **Select subsets of cases.** You can restrict your analysis to a subset of cases or perform simultaneous analyses on different subsets.

The examples in this chapter use the data file *demo.sav*.

Sorting Data

Sorting cases (sorting rows of the data file) is often useful and sometimes necessary for certain types of analysis.

To reorder the sequence of cases in the data file based on the value of one or more sorting variables:

▶ From the menus choose:
Data
   Sort Cases...

Chapter 10

This opens the Sort Cases dialog box.

Figure 10-1
Sort Cases dialog box

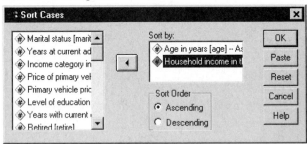

▶ Add the *Age in years (age)* and *Household income in thousands (income)* variables to the Sort By list.

If you select multiple sort variables, the order in which they appear on the Sort By list determines the order in which cases are sorted. In this example, based on the entries in the Sort By list, cases will be sorted by the value of *Household income in thousands (income)* within categories of *Age in years (age)*. For string variables, uppercase letters precede their lowercase counterparts in sort order (for example, the string value *Yes* comes before *yes* in the sort order).

Split-File Processing

To split your data file into separate groups for analysis:

▶ From the menus choose:
Data
   Split File...

Sorting and Selecting Data

This opens the Split File dialog box.

Figure 10-2
Split File dialog box

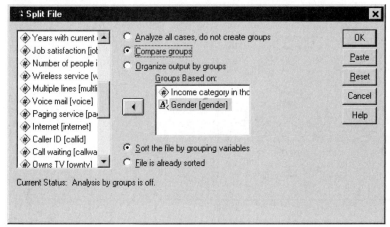

▶ Select Compare groups or Organize output by groups. The examples following these steps show the differences between these two options.

▶ Select *Gender (gender)* to split the file into separate groups for these variables.

You can use numeric, short string, and long string variables as grouping variables. A separate analysis is performed for each subgroup defined by the grouping variables. If you select multiple grouping variables, the order in which they appear on the Groups Based On list determines the manner in which cases are grouped.

Chapter 10

If you select Compare groups and run the Frequencies procedure, a single pivot table is created.

Figure 10-3
Split File output with single pivot table

Statistics

Household income in thousands

Female	N	Valid	3179
		Missing	0
	Mean		68.7798
	Median		44.0000
	Std. Deviation		75.73510
Male	N	Valid	3221
		Missing	0
	Mean		70.1608
	Median		45.0000
	Std. Deviation		81.56216

If you select Organize output by groups and run the Frequencies procedure, two pivot tables are created: one for females and one for males.

Figure 10-4
Split File output with pivot table for females

Statistics[a]

Household income in thousands

N	Valid	3179
	Missing	0
Mean		68.7798
Median		44.0000
Std. Deviation		75.73510

a. Gender = Female

Sorting and Selecting Data

Figure 10-5
Split File output with pivot table for males

Statistics^a

Household income in thousands

N	Valid	3221
	Missing	0
Mean		70.1608
Median		45.0000
Std. Deviation		81.56216

a. Gender = Male

Sorting Cases for Split-File Processing

The Split File procedure creates a new subgroup each time it encounters a different value for one of the grouping variables. Therefore, it is important to sort cases based on the values of the grouping variables before invoking split-file processing.

By default, Split File automatically sorts the data file based on the values of the grouping variables. If the file is already sorted in the proper order, you can save processing time if you select **File is already sorted**.

Turning Split-File Processing On and Off

Once you invoke split-file processing, it remains in effect for the rest of the session unless you turn it off.

- **Analyze all cases.** Turns split-file processing off.
- **Compare groups** and **Organize output by groups.** Turns split-file processing on.

If split-file processing is in effect, the message **Split File on** appears on the status bar at the bottom of the application window.

Chapter 10

Selecting Subsets of Cases

You can restrict your analysis to a specific subgroup based on criteria that include variables and complex expressions. You can also select a random sample of cases. The criteria used to define a subgroup can include:

- Variable values and ranges
- Date and time ranges
- Case (row) numbers
- Arithmetic expressions
- Logical expressions
- Functions

To select a subset of cases for analysis:

▶ From the menus choose:
Data
    Select Cases...

This opens the Select Cases dialog box.

Figure 10-6
Select Cases dialog box

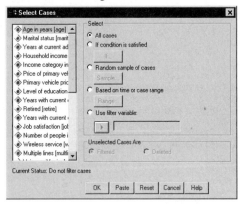

Sorting and Selecting Data

Selecting Cases Based on Conditional Expressions

To select cases based on a conditional expression:

▶ Select If condition is satisfied and click If in the Select Cases dialog box.

This opens the Select Cases If dialog box.

Figure 10-7
Select Cases If dialog box

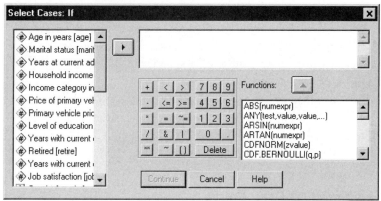

The conditional expression can use existing variable names, constants, arithmetic operators, logical operators, relational operators, and functions. You can type and edit the expression in the text box just like text in an output window. You can also use the calculator pad, variable list, and function list to paste elements into the expression. For more information, see "Using Conditional Expressions" in Chapter 9 on page 162.

Selecting a Random Sample

To obtain a random sample:

▶ Select Random sample of cases in the Select Cases dialog box.

▶ Click Sample.

This opens the Select Cases Random Sample dialog box.

Chapter 10

Figure 10-8
Select Cases Random Sample dialog box

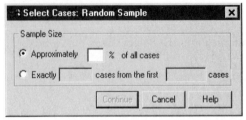

You can select one of the following alternatives for sample size:

- **Approximately.** A user-specified percentage. This option generates a random sample of approximately the specified percentage of cases.

- **Exactly.** A user-specified number of cases. You must also specify the number of cases from which to generate the sample. This second number should be less than or equal to the total number of cases in the data file. If the number exceeds the total number of cases in the data file, the sample will contain proportionally fewer cases than the requested number.

Selecting a Time Range or Case Range

To select a range of cases based on dates, times, or observation (row) numbers:

▶ Select Based on time or case range and click Range in the Select Cases dialog box.

This opens the Select Cases Range dialog box, in which you can select a range of observation (row) numbers.

Sorting and Selecting Data

Figure 10-9
Select Cases Range dialog box

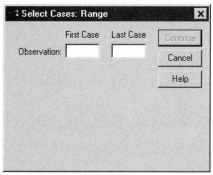

- **First Case.** Enter the starting date and/or time values for the range. If no date variables are defined, enter the starting observation number (row number in the Data Editor, unless Split File is on). If you do not specify a Last Case value, all cases from the starting date/time to the end of the time series are selected.

- **Last Case.** Enter the ending date and/or time values for the range. If no date variables are defined, enter the ending observation number (row number in the Data Editor, unless Split File is on). If you do not specify a First Case value, all cases from the beginning of the time series up to the ending date/time are selected.

For time series data with defined date variables, you can select a range of dates and/or times based on the defined date variables. Each case represents observations at a different time, and the file is sorted in chronological order.

Figure 10-10
Select Cases Range dialog box (time series)

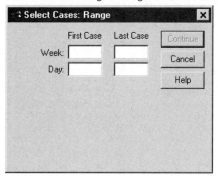

To generate date variables for time series data:

▶ From the menus choose:
Data
  Define Dates...

Unselected Cases

You can choose one of the following alternatives for the treatment of unselected cases:

- **Filtered**. Unselected cases are not included in the analysis but remain in the data file. You can use the unselected cases later in the session if you turn filtering off. If you select a random sample or if you select cases based on a conditional expression, this generates a variable named *filter_$* with a value of 1 for selected cases and a value of 0 for unselected cases.

- **Deleted**. Unselected cases are deleted from the data file. By reducing the number of cases in the open data file, you can save processing time. Deleted cases can be recovered only by exiting from the file without saving any changes and then reopening the file. The deletion of cases is permanent if you save the changes to the data file. *Note*: If you delete unselected cases and save the file, the cases cannot be recovered.

Case Selection Status

If you have selected a subset of cases but have not discarded unselected cases, unselected cases are marked in the Data Editor with a diagonal line through the row number.

Sorting and Selecting Data

Figure 10-11
Case selection status

	age	marital	address	income	inccat	car
1	46	Unmarried	26	24.00	Under $25	12.20
2	34	Unmarried	10	20.00	Under $25	10.00
3	21	Unmarried	0	13.00	Under $25	6.30
4	24	Unmarried	5	24.00	Under $25	11.80
5	21	Unmarried	0	22.00	Under $25	11.00
6	38	Married	0	19.00	Under $25	9.20
7	30	Married	0	20.00	Under $25	9.70
8	22	Unmarried	3	23.00	Under $25	11.50
9	34	Unmarried	15	23.00	Under $25	11.40
10	70	Unmarried	23	21.00	Under $25	10.30
11	48	Married	19	23.00	Under $25	11.40
12	31	Unmarried	4	22.00	Under $25	11.20
13	28	Married	7	15.00	Under $25	7.50
14	50	Unmarried	3	23.00	Under $25	11.70

Chapter

11

Additional Statistical Procedures

This chapter contains brief examples for selected statistical procedures. The procedures are grouped according to the order in which they appear on the Analyze menu.

The examples are designed to illustrate sample specifications required to run a statistical procedure. The examples in this chapter use the data file *demo.sav*, except for the following:

- The paired-samples *t* test example uses the data file *dietstudy.sav*, which is a hypothetical data file containing the results of a study of the "Stillman diet." In the examples in this chapter, you must run the procedures to see the output.

- The correlation examples use *Employee data.sav*, which contains historical data about a company's employees.

- The exponential smoothing example uses the data file *inventor.sav*, which contains inventory data collected over a period of 70 days.

For information about individual items in a dialog box, click **Help**. If you want to locate a specific statistic, such as percentiles, use the Index or Search facility in the Help system. For additional information about interpreting the results obtained by running these procedures, consult a statistics or data analysis textbook.

Summarizing Data

The Descriptive Statistics submenu on the Analyze menu provides techniques for summarizing data with statistics and charts.

Chapter 11

Frequencies

In the chapter *Examining Summary Statistics for Individual Variables*, there is an example showing a frequency table and a bar chart. In that example, the Frequencies procedure was used to analyze the variables *Owns PDA (ownpda)* and *Owns TV (owntv)*, both of which are categorical variables having only two values. If the variable that you want to analyze is a scale (interval, ratio) variable, you can use the Frequencies procedure to generate summary statistics and a histogram. A **histogram** is a chart that shows the number of cases in each of several groups. This example will use Frequencies to analyze the variable *Years with current employer (employ)*.

To generate statistics and a histogram of the years with current employer, follow these steps:

▶ From the menus choose:
Analyze
  Descriptive Statistics
    Frequencies...

This opens the Frequencies dialog box.

Figure 11-1
Frequencies dialog box

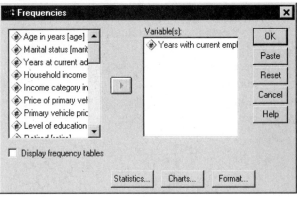

▶ Select *Years with current employer (employ)* and move it to the Variable(s) list.

Additional Statistical Procedures

▶ Deselect the Display frequency tables check box.

(If you leave this item selected and display a frequency table for current salary, the output shows an entry for every distinct value of salary, making a very long table.)

▶ Click Charts to open the Frequencies Charts dialog box.

Figure 11-2
Frequencies Charts dialog box

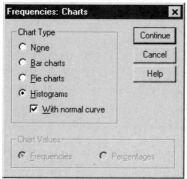

▶ Select Histograms and With normal curve, and then click Continue.

▶ To select summary statistics, click Statistics in the Frequencies dialog box. This displays the Frequencies Statistics dialog box.

Figure 11-3
Frequencies Statistics dialog box

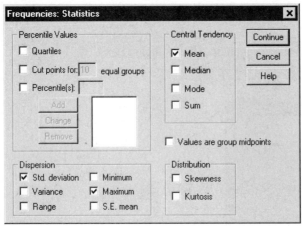

▶ Select Mean, Std. deviation, and Maximum, and then click Continue.

▶ Click OK in the Frequencies dialog box to run the procedure.

The Viewer shows the requested statistics and a histogram in standard graphics format. Each bar in the histogram represents the number of employees within a range of five years, and the year values displayed are the range midpoints. As requested, a normal curve is superimposed on the chart.

Explore

Suppose that you want to look further at the distribution of the years with current employer for each income category. With the Explore procedure, you can examine the distribution of the years with current employer within categories of another variable.

Additional Statistical Procedures

▶ From the menus choose:

Analyze
  Descriptive Statistics
    Explore...

This opens the Explore dialog box.

Figure 11-4
Explore dialog box

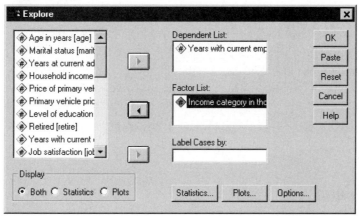

▶ Select *Years with current employer (employ)* and move it to the Dependent List.

▶ Select *Income category in thousands (inccat)* and move it to the Factor List.

▶ Click OK to run the Explore procedure.

In the output, descriptive statistics and a stem-and-leaf plot are displayed for the years with current employer in each income category. The Viewer also contains a boxplot (in standard graphics format) comparing the years with current employer in the income categories. For each category, the boxplot shows the median, interquartile range (25th to 75th percentile), outliers (indicated by O), and extreme values (indicated by *).

More about Summarizing Data

There are many ways to summarize data. For example, to calculate medians or percentiles, use the Frequencies procedure or the Explore procedure. Here are some additional methods:

- **Descriptives.** For income, you can calculate standard scores, sometimes called z scores. Use the Descriptives procedure and select Save standardized values as variables.
- **Crosstabs.** You can use the Crosstabs procedure to display the relationship between two or more categorical variables.
- **Summarize procedure.** You can use the Summarize procedure to write to your output window a listing of the actual values of age, gender, and income of the first 25 or 50 cases. To run the Summarize procedure, from the menus choose:

Analyze
  Reports
    Case Summaries...

Comparing Means

The Compare Means submenu on the Analyze menu provides techniques for displaying descriptive statistics and testing whether differences are significant between two means for both independent and paired samples. You can also test whether differences are significant among more than two independent means by using the One-Way ANOVA procedure.

Additional Statistical Procedures

Means

In the *demo.sav* file, several variables are available for dividing people into groups. You can then calculate various statistics in order to compare the groups. For example, you can compute the average (mean) household income for males and females. To calculate the means, use the following steps:

▶ From the menus choose:
Analyze
  Compare Means
    Means...

This opens the Means dialog box.

Figure 11-5
Means dialog box (layer 1)

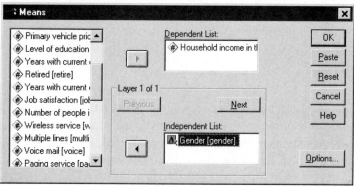

▶ Select *Household income in thousands (income)* and move it to the Dependent List.

▶ Select *Gender (gender)* and move it to the Independent List in layer 1.

▶ Click Next. This creates another layer.

Figure 11-6
Means dialog box (layer 2)

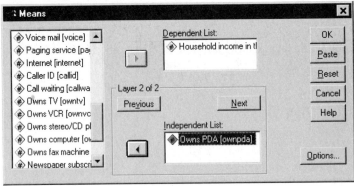

- Select *Owns PDA (ownpda)* and move it to the Independent List in layer 2.

- Click OK to run the procedure.

Paired-Samples T Test

When the data are structured in such a way that there are two observations on the same individual or observations that are matched by another variable on two individuals (twins, for example), the samples are paired. In the data file *dietstudy.sav*, the beginning and final weights are provided for each person who participated in the study. If the diet worked, we expect that the participant's weight before and after the study would be significantly different.

To carry out a *t* test of the beginning and final weights, use the following steps:

- Open the data file *dietstudy.sav*, which can be found in the *tutorial**sample_files*\\ subdirectory of the directory in which you installed SPSS.

▶ From the menus choose:
Analyze
  Compare Means
    Paired-Samples T Test...

This opens the Paired-Samples T Test dialog box.

Figure 11-7
Paired-Samples T Test dialog box

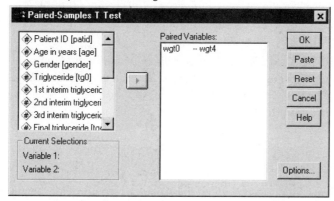

▶ Click *Weight (wgt0)*. The variable is displayed in the Current Selections group (below the variable list).

▶ Click *Final weight (wgt4)*. The variable is displayed in the Current Selections group.

▶ Click the arrow button to move the pair to the Paired Variables list.

▶ Click OK to run the procedure. If there are rows of asterisks in some columns, double-click the chart and drag to make the columns wider.

The results show that the final weight is significantly different from the beginning weight, as indicated by the small probability displayed in the *Sig. (2-tailed)* column of the Paired Samples Test table.

More about Comparing Means

The following examples suggest some ways in which you can use other procedures to compare means.

- **Independent-Samples T Test.** When you use a *t* test to compare means of one variable across independent groups, the samples are independent. Males and females in the *demo.sav* file can be divided into independent groups by the variable *Gender (gender)*. You can use a *t* test to determine if the mean household incomes of males and females are the same.
- **One-Sample T Test.** You can test whether the household income of people with college degrees differs from a national or state average. Use Select Cases on the Data menu to select the cases with *Level of Education (ed)* >= 4. Then, run the One-Sample T Test procedure to compare *Household income in thousands (income)* and the test value 75.
- **One-Way ANOVA.** The variable *Level of Education (ed)* divides employees into five independent groups by level of education. You can use the One-Way ANOVA procedure to test whether *Household income in thousands (income)* means for the five groups are significantly different.

ANOVA Models

The General Linear Model submenu on the Analyze menu provides techniques for testing univariate analysis-of-variance models. (If you have only one factor, you can use the One-Way ANOVA procedure on the Compare Means submenu.)

Univariate Analysis of Variance

The GLM Univariate procedure can perform an analysis of variance for factorial designs. A simple factorial design can be used to test if a person's household income and job satisfaction affect the number of years with current employer.

▶ From the menus choose:
Analyze
  General Linear Model
    Univariate...

This opens the Univariate dialog box.

Figure 11-8
Univariate dialog box

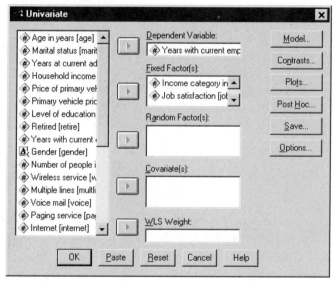

▶ Select *Years with current employer (employ)* as the dependent variable.

▶ Select *Income category in thousands (inccat)* and *Job satisfaction (jobsat)* as fixed factors.

▶ Click OK to run the procedure.

In the Tests of Between-Subjects Effects table, you can see that the effects of income and job satisfaction are definitely significant and that the observed significance level of the interaction of income and job satisfaction is 0.000. For further interpretation, consult a statistics or data analysis textbook.

Chapter 11

Correlating Variables

The Correlate submenu on the Analyze menu provides measures of association for two or more numeric variables.

The examples in this topic use the data file *Employee data.sav*.

Bivariate Correlations

The Bivariate Correlations procedure computes statistics such as Pearson's correlation coefficient. Correlations measure how variables or rank orders are related. Correlation coefficients range in value from −1 (a perfect negative relationship) and +1 (a perfect positive relationship). A value of 0 indicates no linear relationship.

For example, you can use Pearson's correlation coefficient to see if there is a strong linear association between *Current Salary (salary)* and *Beginning Salary (salbegin)* in the data file *Employee data.sav*.

Partial Correlations

The Partial Correlations procedure calculates partial correlation coefficients that describe the relationship between two variables while adjusting for the effects of one or more additional variables.

You can estimate the correlation between *Current Salary (salary)* and *Beginning Salary (salbegin)*, controlling for the linear effects of *Months since Hire (jobtime)* and *Previous Experience (prevexp)*. The number of control variables determines the order of the partial correlation coefficient.

To carry out this Partial Correlations procedure, use the following steps:

▶ Open the *Employee data.sav* file. It is usually in the directory where SPSS is installed.

Additional Statistical Procedures

▶ From the menus choose:
Analyze
  Correlate
    Partial...

This opens the Partial Correlations dialog box.

Figure 11-9
Partial Correlations dialog box

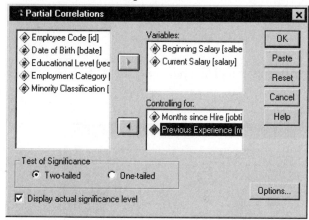

▶ Select *Current Salary (salary)* and *Beginning Salary (salbegin)* and move them to the Variables list.

▶ Select *Months since Hire (jobtime)* and *Previous Experience (prevexp)* and move them to the Controlling For list.

▶ Click OK to run the procedure.

The output shows a table of partial correlation coefficients, the degrees of freedom, and the significance level for the pair *Current Salary (salary)* and *Beginning Salary (salbegin)*.

Regression Analysis

The Regression submenu on the Analyze menu provides regression techniques.

Chapter 11

Linear Regression

The Linear Regression procedure examines the relationship between a dependent variable and a set of independent variables. You can use it to predict a person's household income (the dependent variable) from independent variables such as age, number in household, and years with employer.

▶ From the menus choose:
Analyze
  Regression
    Linear...

This opens the Linear Regression dialog box.

Figure 11-10
Linear Regression dialog box

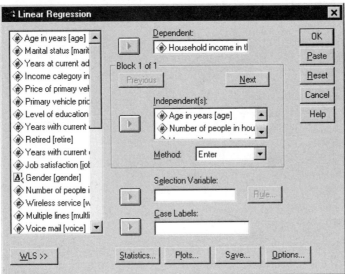

▶ Select *Household income in thousands (income)* and move it to the Dependent list.

▶ Select *Age in years (age)*, *Number of people in household (reside)*, and *Years with current employer (employ)* and move them to the Independent(s) list.

▶ Click OK to run the procedure.

Additional Statistical Procedures

The output contains goodness-of-fit statistics and the partial regression coefficients for the variables.

Examining fit. To see how well the regression model fits your data, you can examine the residuals and other types of diagnostics that this procedure provides. In the Linear Regression dialog box, click **Save** to see a list of the new variables that you can add to your data file. If you generate any of these variables, they will not be available in a later session unless you save the data file.

Methods. If you have collected a large number of independent variables and want to build a regression model that includes only variables that are statistically related to the dependent variable, you can select a method from the drop-down list. For example, if you select **Stepwise** in the above example, only variables that meet the criteria in the Linear Regression Options dialog box are entered in the equation.

Nonparametric Tests

The Nonparametric Tests submenu on the Analyze menu provides nonparametric tests for one sample or for two or more paired or independent samples. Nonparametric tests do not require assumptions about the shape of the distributions from which the data originate.

Chi-Square

The Chi-Square Test procedure is used to test hypotheses about the relative proportion of cases falling into several mutually exclusive groups. You can test the hypothesis that people who participated in the survey occur in the same proportions of gender as the general population (50% males, 50% females).

In this example, you will need to recode the string variable *Gender (gender)* into a numeric variable before you can run the procedure.

▶ From the menus choose:
Transform
  Automatic Recode...

This opens the Automatic Recode dialog box.

Figure 11-11
Automatic Recode dialog box

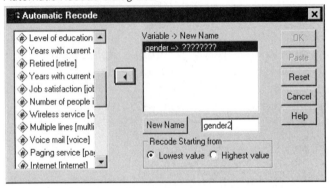

▶ Select the variable *Gender (gender)* and move it to the Variable -> New Name list.

▶ Type gender2 in the New Name text box, and then click the New Name button.

▶ Click OK to run the procedure.

This creates a new numeric variable called *gender2*, which has a value of 1 for females and a value of 2 for males. Now a chi-square test can be run with a numeric variable.

▶ From the menus choose:
Analyze
  Nonparametric Tests
    Chi-Square...

▶ This opens the Chi-Square Test dialog box.

Figure 11-12
Chi-Square Test dialog box

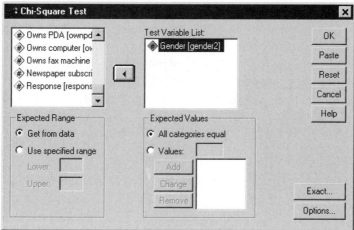

- Select *Gender (gender2)* as the test variable.

- Select All categories equal, since, in the general population of working age, the number of males and females is approximately equal.

- Click OK to run the procedure.

    The output shows a table of the expected and residual values for the categories. The significance of the chi-square test is 0.6. Consult a statistics or data analysis text book for more information on interpretation of the statistics.

Time Series Analysis

A time series variable is a variable whose values are recorded at regular intervals over a period of time. The Time Series submenu on the Analyze menu provides exponential smoothing that can be used for predictions. Exponential Smoothing is available in the Student Version and in the Trends module.

Exponential Smoothing

The Exponential Smoothing procedure performs exponential smoothing of time series data. It creates new series containing predicted values and residuals. This topic uses the *Inventor.sav* file.

For example, you can fit a model for inventory data and use it to predict the next week's inventory. Suppose that for 70 days you have kept track of the inventory of power supplies and that you want to construct a model and then use it to forecast power supplies for the next week.

To use the Exponential Smoothing procedure to forecast power supplies, follow these steps:

▶ Open the *Inventor.sav* file. It is usually in the directory where SPSS is installed.

▶ From the menus choose:
Analyze
    Time Series
        Exponential Smoothing...

This opens the Exponential Smoothing dialog box.

Figure 11-13
Exponential Smoothing dialog box

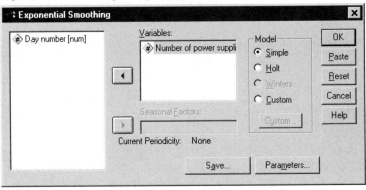

▶ Select *Number of power supplies (amount)* and move it to the Variables list.

Additional Statistical Procedures

▶ Click Parameters to specify the procedure.

This opens the Exponential Smoothing Parameters dialog box

Figure 11-14
Exponential Smoothing Parameters dialog box

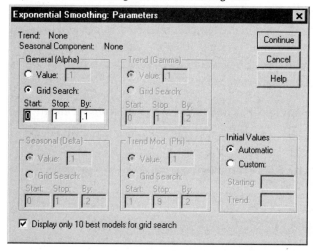

▶ To search for the best general parameter, select Grid Search and then click Continue.

▶ To create a new variable that contains predicted values, click Save.

This opens the Exponential Smoothing Save dialog box.

Figure 11-15
Exponential Smoothing Save dialog box

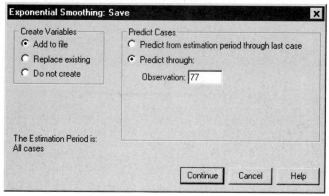

▶ Select Predict through and type 77 in the Observation text box.

This adds 7 days to the original 70.

▶ Click Continue and then in the main dialog box, click OK.

This runs the procedure and adds the new variables *fit_1* and *err_1*. The variable *fit_1* contains the fitted values and the seven new predicted values. The variable *err_1* contains residual values for the original 70 cases; you can use the residuals for further analysis. If you want to save the new variables, select Save As from the File menu and save the data file under a new name.

▶ To see a chart of the original data and the new fit line, from the menus choose:
Graphs
  Sequence...

This opens the Sequence Charts dialog box.

Figure 11-16
Sequence Charts dialog box

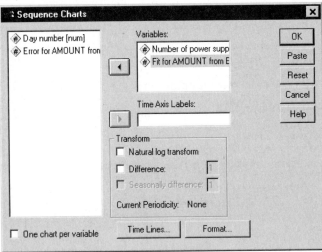

▶ Select *Number of power supplies (amount)* and *Fit for AMOUNT (fit_1)* and move them to the Variables list.

▶ Click OK to run the procedure.

The resulting chart shows both the actual number of power supplies and the fit line plotted on the same axes. The predicted values are plotted for the next week on the right side of the chart.

Index

Access (Microsoft), 34

bar charts, 81

cases
    selecting, 174
    sorting, 169, 173
case studies, 27
categorical data, 79
    summary measures, 80
charts
    bar, 81
    creating charts from pivot tables, 143
    creating interactive charts, 135
    creating standard charts, 125
    editing interactive charts, 139
    editing standard charts, 128
    histograms, 86
computing new variables, 159
conditional expressions, 162
continuous data, 79
copy value attributes, 66
counts
    tables of counts, 80
create variable labels, 56

database files
    reading, 34
Database Wizard, 34
Data Editor
    entering non-numeric data, 54
    entering numeric data, 51
data entry, 51, 54
data types
    for variables, 57

editing pivot tables, 98
entering data
    non-numeric, 54
    numeric, 51
Excel (Microsoft)
    exporting results to, 117
Excel files
    reading, 31

exporting results
    to Excel, 117
    to HTML, 123
    to text, 123
    to Word, 117

finding Help topics, 17
frequency tables, 80
functions in expressions, 161

graphs
    creating graphs from pivot tables, 143
    creating interactive graphs, 135
    creating standard graphs, 125
    editing interactive graphs, 139
    editing standard graphs, 128

Help
    case studies, 27
    Statistics Coach, 19
Help windows
    contents, 16
    index, 17
hiding rows and columns in pivot tables, 100
histograms, 86
HTML
    exporting results to, 123

index Help topic search, 17
interactive charts, 135, 139
interval data, 79

keyword Help topic search, 17

layers
    creating in pivot tables, 96
level of measurement, 79

measurement level, 79
missing values
    for non-numeric variables, 65
    for numeric variables, 63
    system missing, 62
moving
    elements in pivot tables, 93
    items in the Viewer, 89

nominal data, 79

Index

numeric data, 51

ordinal data, 79

pasting syntax
    from a dialog box, 147
pivot tables
    accessing definitions, 92
    cell data types, 100
    cell formats, 100
    editing, 98
    formatting, 98
    hiding decimal points, 100
    hiding rows and columns, 100
    layers, 96
    pivoting trays, 93
    transposing rows and columns, 93

qualitative data, 79
quantitative data, 79

ratio data, 79
recoding values, 153

scale data, 79
scale variables
    summary measures, 83
selecting cases, 174
sorting cases, 169
split-file processing, 170
spreadsheet files
    reading, 31
    reading variable names, 31
SPSS
    starting, 2
Statistics Coach, 19
string data
    entering data, 54
subsets of cases
    based on dates and times, 176
    conditional expressions, 175
    deleting unselected cases, 178
    filtering unselected cases, 178
    if condition is satisfied, 175
    random sample, 175
    selecting, 174
summary measures
    categorical data, 80
    scale variables, 83
syntax, 147
syntax files
    opening, 152
    saving, 152
Syntax Help tool, 150
syntax windows
    editing commands, 150
    opening a new window, 151
    pasting commands, 147
    running commands, 149, 152
system-missing values, 62

text data files
    reading, 40
Text Import Wizard, 40
transposing (flipping) rows and columns in pivot
        tables, 93

value attributes
    reuse, 66
value labels
    assigning, 58, 60
    controlling display in Viewer, 58, 60
    non-numeric variables, 60
    numeric variables, 58
variable labels
    creating, 56
variables, 51
    data types, 57
    labels, 56
Viewer
    hiding and showing output, 89
    moving output, 89

Word (Microsoft)
    exporting results to, 117